CAMBRIDGE COUNTY GEOGRAPHIES

General Editor: F. H. H. Guillemard, M.A., M.D.

SURREY

Cambridge County Geographies

SURREY

by

GEORGE F. BOSWORTH, F.R.G.S.

With Maps, Diagrams and Illustrations

Cambridge:
at the University Press
1909

CAMBRIDGE UNIVERSITY PRESS
Cambridge, New York, Melbourne, Madrid, Cape Town,
Singapore, São Paulo, Delhi, Mexico City

Cambridge University Press
The Edinburgh Building, Cambridge CB2 8RU, UK

Published in the United States of America by
Cambridge University Press, New York

www.cambridge.org
Information on this title: www.cambridge.org/9781107675988

First published 1909
First paperback edition 2012

A catalogue record for this publication is available from the British Library

ISBN 978-1-107-67598-8 Paperback

CONTENTS

CONTENTS

ILLUSTRATIONS

MAPS

The illustrations on pp. 11, 15, 20, 24, 64, 67, 70, 75, 77, 89, 91, 95, 98, 101, 103, 106, 108, 119 and 136 are from photographs by The Homeland Association, Ltd.; and those on pp. 29, 31, 44, 62, 122 and 125 are from photographs by Messrs F. Frith & Co., Ltd., Reigate.

1. The South-Eastern Peninsula. County and Shire. The Word *Surrey*. Its Origin and Meaning.

The three counties of Kent, Surrey, and Sussex form a peninsula in the south-east of England, having the Thames on the north, and the sea on the east and south, while the boundary on the west is formed by Berkshire and Hampshire. This south-eastern peninsula has always been of the greatest importance in our history, for the chief lines of communication between the Continent and London pass through one or other of the three counties. The Thames is the waterway to London; the old roads from Dover, Hastings, and Portsmouth are the highways through this peninsula to the metropolis; and the modern railways from the chief sea-ports of the south-east carry passengers and goods to the great city.

It will now be understood that each of these three south-eastern counties is of considerable importance on account of the proximity of London, and it is both interesting and instructive to have a definite knowledge of the past and present condition of all of them. In this book we are concerned with Surrey, one of the "home counties" as it is called; and it will be well for us at the outset to find out what is meant by a county and then discover how Surrey came by its name.

Have you ever thought how the present division of England into counties was brought about? We often look at the map and notice that some of the counties are

Wharves on the Surrey Side

large and some are small; and we notice, too, that some of the names end in *-shire.* If we do this, we must enquire further, as to how this difference in size and name happens. Some books tell us that King Alfred divided England into counties. Now this is quite wrong, for although that great king did many things, he certainly did not accomplish the division of England into counties. We know that some of the counties existed before his time, and that others were not formed till after his death.

The fact is that some of our present counties are survivals of the old English kingdoms, which have kept, in many cases, their former names and extent. Kent, Surrey, and Sussex are such. Others of our counties are *shires*, or *shares* of former large kingdoms, such as Mercia, or Wessex, or Northumbria. For instance, Mercia was a large English kingdom in the middle of our land, and at various times portions of it were cut off, and we now know these divisions as Stafford*shire*, Warwick*shire*, and so on.

Thus we may say quite correctly that our English counties have grown, and it is this gradual growth that makes their history so interesting. When we trace out their boundaries, we find that they generally represent the limits of tribes or kingdoms. Some of them represent the possessions of ancient British tribes, such as Kent, which was the land of the Cantii; some of them—such as Sussex, and, perhaps, Surrey—are the representatives of early English kingdoms; and others were constituted by the Danes, or by the Conqueror at a later date.

The change is still going on, and towards the end of the nineteenth century several of our counties were divided for administrative purposes, while portions of Kent, Surrey, and Sussex were detached from the ancient counties and added to London, so as to form the modern County of London.

We shall understand our geography and history so

much better if we grasp this fact clearly, that many of our counties are the same, or nearly the same, as the first English kingdoms, which were never less than seven in number, and often far more numerous. Thus Essex was the land of the East Saxons, and Sussex of the South Saxons. Essex is a county marked off by sea and rivers from the adjoining land, and Sussex was a district naturally marked off by sea, and woods, and marshes from its neighbours. Surrey, however, differs in name and origin from these counties. It is quite easy to discover the meaning of Essex and Sussex, but the origin of the word Surrey is much more difficult to ascertain. Surrey, or Suthrege, is probably "the south kingdom." The first part of the word means south, while *rege* comes from the Anglo-Saxon *ric*, a kingdom, the same termination that occurs in bishopric. In some early charters Surrey is called *Sudergeona*, and *Sudregona*. From this it is evident that Surrey was in early days an independent kingdom, or sub-kingdom, and its name was given with reference to the position of Middlesex or Essex.

When we first hear of Surrey, it was ruled by Essex; and after a time it passed, with the rest of the southern counties of England, under the sway of Wessex. There is thus some obscurity as to the way in which Surrey came to rank as a shire or county. So little is known about its first settlement, that we can only guess as to its early organisation. But we shall not be far wrong if we say that Surrey is a shire that had its origin in an early kingdom.

2. General Characteristics. Its Position and Natural Conditions.

Surrey is entirely an inland county. Its only connection with the sea arises from the fact that the tide runs up the Thames as far as Teddington. Thus from a commercial point of view it has not the natural advantages of either Kent or Sussex.

The importance of Surrey is not, however, to be estimated by the same standard as that we apply to either of its neighbouring counties. It is not so large as Kent or Sussex; indeed it is a small county. Its population is denser than that of most English counties, and it has a good number of towns of more or less importance.

Surrey undoubtedly owes its importance to its nearness to the metropolis, and to the good railway facilities, which give access to all parts of this delightful county. Thus we find that successful London merchants and men of all professions make it their home. Artists, poets, and men of science find in Surrey a beautiful resort where they can quickly escape from the noise and bustle of London.

It would be difficult to find a county in any part of England that offers so many attractions as Surrey. Its climate, on the whole, is dry, and in the higher parts distinctly bracing. The commons and heathlands are charming at all times, particularly in the spring when the gorse and may are in bloom. The south-west corner is as beautiful as any part of our island. Indeed the district

TEDDINGTON·LOCK

around Hindhead has been called the miniature Highlands of England.

No better testimony to the attractions of Surrey can be given than the fact that it draws vast crowds of cyclists and pedestrians to its hills and dales, its parks and woods, and its glorious commons. In the seventeenth century the great diarist and nature-lover John Evelyn boasted that Surrey was the county of his birth and his delight; and since his day its praises have been sung by many of our greatest poets. Tennyson was so fascinated with the beauty of the county that he built Aldworth House near Haslemere, where he spent the closing years of his life.

The north-eastern part of the county is rapidly becoming a part of London. Such towns as Wimbledon and Croydon, Surbiton and Sutton are losing their rural character, and houses are being built to accommodate the workers of London. It seems a pity that this once beautiful district should now be covered with thousands of houses of the most monotonous character. But that is part of the price that it has to pay for its proximity to London. Surrey is one of the residential counties of the metropolis, and is fast losing its character as an agricultural county. Thus it came about that Surrey, in its earliest days, owed its formation to the existence of London in the north-east; and it now still gains its wealth and importance from the merchants and workers of our capital.

Surrey will always be famous as the county which is associated with the annual boat-race between Oxford and Cambridge Universities. Perhaps no stretch of the

A GLIMPSE of HIND HEAD

Thames is better known than the famous course between Putney and Mortlake, where, every year, this classic race is witnessed by tens of thousands of Englishmen. Surrey, too, has been distinguished as the county that has favoured the national game of cricket. Every cricketer knows the "Oval," where Surrey cricketers have won fame for many years. Not only at the Oval, but on many a village green and Surrey field has cricket been played with advantage to all concerned. Epsom is the place in Surrey that is associated with the sport of horse-racing, which every year attracts countless throngs of spectators of the most varied type and character.

3. Size. Shape. Boundaries.

We have seen in the previous chapters how Surrey came to be a county and we have learnt something of its general characteristics. Now let us consider the size and shape of it, and we must look at a good map to trace its boundaries. Before we begin this chapter, we must remember that the ancient county of Surrey was larger than the present county, and so it will be necessary to speak of Surrey, sometimes as including the portion that is now comprised in the County of London, and sometimes as excluding this portion. Whenever we refer to the larger Surrey, we shall speak of it as the Ancient County, and when we refer to the smaller Surrey, we shall speak of it as the Administrative County.

Surrey is an inland county lying along the south of

the river Thames. From east to west its greatest extent is about forty miles, and from north to south it is about twenty-four miles.

The area of the ancient county is 485,122 acres, or about 758 square miles. The area of the administrative county is 461,829 acres, or about 722 square miles. If we take the larger area, we shall find that Surrey is about one half the area of Kent, or nearly the same size as Berkshire. There are ten counties in England smaller than Surrey, so that it is one of the smallest of the English counties. It occupies about one sixty-seventh of the entire area of England and Wales.

Except on the north side the area of Surrey is comprised within fairly regular outlines. The Thames forms a very irregular northern boundary, but, roughly speaking, the shape of Surrey is quadrilateral. This is more especially the case if we speak of the ancient county. In 1889, a large and populous portion in the north-east was taken away and added to London for administrative purposes. But even in speaking of the administrative county of Surrey, we shall not be far wrong in saying that it is somewhat the shape of an oblong.

When we come to deal with the boundaries of a county, we have always some interesting questions to answer. It is very easy to state the present boundaries, but it is not always so easy to say how these boundaries were settled. However, we will try, and with the aid of a good map and a knowledge of our early history, we may get a fairly accurate idea.

The Thames is the northern boundary of the ancient

Runnymede

county from Deptford to Runnymede. This is practically the only natural boundary of Surrey, and it is easy to see that it was the best of all boundaries when Surrey, the south kingdom, was separated from Essex on the north side of the Thames. For we must remember that in those far-off days Essex, the land of the East Saxons, also included Middlesex; so that Surrey was indeed the south kingdom to Essex.

The boundary on the north-west is an artificial one, and separates Surrey from Berkshire. This line runs from Runnymede to York Town, and may have been an irregular boundary along the high land that prevails in this part, passing through a forest of which we can now trace the remains in Windsor Great Park. From York Town there is a little river, the Blackwater, that forms the boundary between Surrey and Hampshire for a distance of about eight miles from Tongham to Chobham. Thence the boundary between the two counties continues irregularly and artificially to Haslemere in the south-west.

Now when we trace the boundary between Surrey and Sussex, we find the line is much more regular. From Haslemere to the borders of Kent, the boundary line runs through the midst of the district known as the Weald. It will be quite fair to say that the South Saxons, who dwelt in Sussex, probably pushed their way northward from their settlements along the coast and, reaching this high ridge, found that the Saxons from the north were in possession of the land north of this ridge. Thus it would be a natural thing for the two tribes to adopt this as the boundary between their settlements.

On the east, Kent forms the boundary, but the line of separation is somewhat irregular and artificial. How this boundary was formed it would take too long to discuss, but we may be sure that the Kentish kingdom had pushed itself as far west as it wished to go, and beyond that limit Surrey was allowed to exist.

We can now see that Surrey was something of a buffer state placed between the East and Middle Saxons on the north and the South Saxons on the south; and between the West Saxons on the one side and the Jutes of Kent on the east.

4. Surface and General Features.

The surface of Surrey is very varied, for while it has much low-lying land, it has some of the highest hills in the south-east of England. A glance at the map will show us that the surfaces of Kent, Surrey, and Sussex have many points in common. Each county is intersected by a range of hills that runs from west to east; the North Downs slope northwards in Kent and Surrey to the Thames; and each county has within its borders the remains of the ancient forest known as Andred's-Wald.

We shall find, however, that while each county has its rounded downs and quiet peaceful valleys, Surrey has some surface features that are wanting in Kent and Sussex. The south-western portion of Surrey is quite unlike any other part of the three south-eastern counties, and for diversity of surface and beauty of scenery, it stands well-nigh unrivalled in England.

The great feature of the surface of Surrey is the range of hills known as the North Downs, which traverse the county from west to east, and then pass into Kent. They enter the county from Hampshire by a ridge known as the Hog's Back, which extends for a distance of nearly ten miles from Farnham to Guildford. This ridge is narrow, but eastward from Guildford the downs broaden out considerably. To the south, the North Downs have a steep slope to the Weald below; while to the north they widen out like a plain, but with a cultivated surface.

The North Downs reach their greatest breadth of ten miles in the east of Surrey, on the borders of Kent. Look at the map and find Leatherhead, Epsom, and Croydon. These places stand along the northern edge of the North Downs. Now find Farnham, Guildford, Dorking, and Tatsfield. These places you will notice are all in a line to the south of the chalk formation, about which you will read in another chapter. The highlands of Surrey are all on the chalk, and you have only to mention the Hog's Back, Box Hill, Epsom Downs, and Banstead Downs to remember that this is the case.

Now you will like to know which are really the highest hills in Surrey, and find where they are situated. Surrey claims the distinction of having the highest hill in the three south-eastern counties. Leith Hill, to the south of Dorking, occupies the premier position, for it reaches a height of 960 feet. The second highest hill in Surrey is Hindhead, and its height is 895 feet. Hindhead is in the extreme south-west of Surrey, a district which was called by Professor Tyndall the "Switzerland of England." The whole district of which Hindhead is the centre is one of great beauty and has attractions that no other part of the county can offer. Just over the Sussex border, this hilly district rises still higher, and Blackdown, the highest hill in Sussex, is 918 feet in height.

Keeper's Cottage, Devil's Punchbowl, Hindhead

Many other hills in Surrey are known owing to their varied attractions. Richmond Hill in the north commands a world-famed view of the valley of the Thames, painted by hundreds of artists from Turner downwards; while Box Hill, near Dorking, is one of the most attractive points. It gains its name from the box tree that flourishes on its slopes. The view from its summit is an extensive one, stretching over a most beautiful country.

It will thus be gathered that the scenery of Surrey is picturesque, varied, and interesting. In the northern portion, the finest points of view are those from Richmond Hill and Cooper's Hill. Along the chalk range extensive and charming views are gained from Box Hill, Leith Hill, St Martha's Chapel, and the Hog's Back. These views afford prospects over a country that is so wild and unspoiled, that one can hardly realise that the great metropolis is not far off.

In the western part of Surrey there are very extensive heaths. From Egham, on the Thames, south-westward to Ash, the district consists mainly of heath and moor, as also does that in a transverse direction from Bagshot by Chobham to Cobham and Oatlands.

5. Watershed. Rivers. The Mole.

We shall understand the river system and drainage of Surrey much better if we remember that this county belongs to a region stretching from Wiltshire to Kent, whose watershed or divide is along the Wealden heights. Now look carefully at the map and you will find that the Wey, Mole, and, in Kent, the Medway flow northwards into the Thames; while the Salisbury Avon, Itchen, Sussex Arun, and Sussex Ouse flow southwards to the sea.

Look again at the map and you will find that all these rivers have cut gaps through the North and South Downs. For instance, the two Surrey rivers, the Wey and the Mole, have their sources in the Wealden heights, and cut through the North Downs on their way to the Thames. It is very important to remember this fact, for these gaps really form the route along which the roads to London must pass. It is also worth noting that these breaks or passes in the downs form the sites of the chief towns of the county. This is specially the case with regard to Farnham, Guildford, Dorking, and Reigate.

Surrey is not famous for its rivers. The Thames borders it on the north, and the Wey and the Mole, which are the principal rivers, both rise in other counties. The Thames is so important that we shall give a separate chapter to it.

The Mole rises in Sussex, but is of little importance till it enters Surrey. It is somewhat difficult to determine which is the head stream of the Mole, but the main branch is that which rises at Rusper, about two miles from Horsham Common, and enters Surrey near Charlwood, a little below which it is joined by another stream from Tilgate Forest.

There is little of importance in the first few miles of its course. After passing Horley, it is joined by a little river from Worth in Sussex, and soon after by a really beautiful stream which rises at the foot of the hill near Merstham Church in Surrey. Thus increased, the Mole leaves the little town of Reigate on the north, and pursues its way past Wonham to Betchworth.

The Mole is now a river of some size, and the scenery through which it flows is exceedingly picturesque. The river passes through Betchworth Park, one of the most beautiful in the county, and then flows on to Dorking, a quiet little town, in whose neighbourhood is the famous Box Hill.

THE MILL COBHAM

About a mile from Dorking, the Mole seems to run into the earth, and after some distance it reappears to flow on as usual. These windings and upturnings are known locally as "swallows," and the fact that the river seemed to burrow underground was often noticed by some of our old poets. Thus Drayton says:—

"Mole digs herself a path, by working day and night,
And underneath the earth for three miles space doth creep,"

and Pope has this line :—

"Sullen Mole, that hides his diving head."

The Mole flows through the vale of Mickleham, after passing Box Hill, and the scenery on both sides of the river is varied and pleasing. The downs are close on either hand, and all along the valley there is a combination of soft swelling downs and cultivated land; fine mansions and magnificent parks; rural churches and picturesque cottages.

Right onward to Leatherhead, the Mole is one of the prettiest rivers. It has many little islets and often reminds one of the Thames in its pleasanter parts. Leatherhead is a small town, which was once famous. Here the Mole widens considerably and is crossed by a fine bridge. From Leatherhead, by Platsome Green to Stoke d'Abernon, the Mole runs through private enclosures. The river now passes by Cobham Park to Esher, a pretty village where Wolsey once had a palace. Thomson speaks of the "soft windings of the silent Mole" at Esher, and although the place is somewhat changed, it is still one of Surrey's pleasant spots.

The Mole separates into two branches at Esher. The one runs by Esher Court and near Thames Ditton; the other flows towards Moulsey Hurst. The banks of the Mole are here low and marshy, and there is little more that is attractive in its course till it joins the Thames nearly opposite Hampton Court.

There is some doubt as to the origin of the word Mole. The river has another name, the Emlyn Stream;

and thus its name may be derived from a word *molins*, a mill, and probably means the mill-stream.

The Mole at Cobham

6. Rivers. The Wey. The Wandle. The Smaller Streams.

The Wey rises in Hampshire, about a mile south-west of Alton. The river flows through the town, but is a very small stream for a long while after leaving it. The valley through which it flows between Alton and Farnham is very fertile and pleasant, indeed Arthur Young calls it "the finest ten miles in England." Cobbett, who was born at Farnham, says, "here is a river with fine meadows on each side, and with rising grounds on each outside of the meadows, those grounds having some hop-gardens and some pretty woods."

The Wey passes just outside Farnham on to Bourne. The course of the river is now very irregular until Godalming is reached. From this town to Guildford there is nothing wild and bold, but the scenery is exceedingly pretty; indeed, Cobbett says "there is hardly another such a pretty four miles in England." Guildford, the county town, has a fine situation, and the river now becomes navigable. It is at this gap in the downs that the Wey makes its way northwards.

A little above Guildford, the Wey receives the Tillingbourne, a stream that drains the northern side of the Leith Hill ridge. The river now runs on to Stoke, a delightful little place, and soon after reaches Woking, where Henry VIII spent some of his early years. Here the Wey divides into several streams, which unite at Wisley. A glance at the map will show that the Wey and the Mole approach within a mile of each other, though they separate again almost directly. The next place of importance on the Wey is Byfleet, and then the river flows on to Weybridge, giving it its name. It enters the Thames not far from Walton, and nearly opposite the Cowey Stakes, where Caesar is supposed to have crossed the Thames on his second expedition.

The Wey takes its name from an old Keltic word meaning water. The names of many rivers are derived from the same root. Thus we have the Wye in Derbyshire, the Con*way* in Wales, the Sol*way* in Scotland, and the Med*way* in Kent.

·THE·KEEP·GARDEN·
·FARNHAM·
·CASTLE·

Besides the Mole and the Wey, the Surrey rivers are few and of little importance. In the north-east of the county is the Wandle, which rises near Croydon. It flows eastward by Beddington and Carshalton, and then northward past Morden, Merton, and Tooting to Wandsworth, to which place it gives its name. Here it enters the Thames after a course of about ten miles.

The Hog's Hill river rises in the town of Ewell, north of Epsom Downs. It pursues its course in a northerly direction and joins the Thames at Kingston. The Beverley Brook rises at Cheam and flows into the Thames near Putney. Both these streams drain the country north of the Epsom Downs.

In the north-west of the county is Bourne Brook, which has its source in the Chobham heights. This stream enters the Thames a little to the east of Chertsey.

Some of the upper streams of the Medway are in the south-east of Surrey; and the Arun, a Sussex river, has its origin in the neighbourhood of Leith Hill. The only other river that is worth mentioning is the Blackwater, a Hampshire river, which, as we have seen, forms part of the Surrey boundary for several miles on the west, between Aldershot and Sandhurst.

7. Rivers. The River Thames. (*a*) To Teddington.

The Thames may be considered as belonging to Surrey from Runnymede to Deptford Creek. Thus Surrey is on its right bank for over 40 miles, and the

Leith Hill Tower

Thames is navigable the whole of this distance. It is a tidal river for more than 20 miles of its course along Surrey, the tide running up to Teddington Lock.

The Thames is the greatest of all our rivers, so we may spend just a short time in considering its source and course before it becomes a Surrey river. It is interesting to note that the Thames, or Tamesis as it was once called, is the earliest British river mentioned in Roman history. Its name is of Keltic origin, and there are several other rivers which have probably the same derivation. Thus we have the Tame, the Teme, and the Tamar in other parts of England.

The Thames rises at Coates in Gloucestershire, not far from the borders of Wiltshire. By the side of a canal that runs through that little village there is an ash tree, encircled with ivy, and in its bark are cut two letters, T.H.—"Thames Head." At the foot of the tree are some stones, indicating that a spring runs under the canal. Near by are other springs, and the water from them flows through the meadows and soon enters the county of Wilts as the Thames. The upper part of the main stream is often called the Isis, and not the Thames, until it has received the waters of the Thame near Dorchester in Oxfordshire.

The Thames has Oxfordshire, Buckinghamshire, Middlesex, and Essex on its left bank; and Wiltshire, Berkshire, Surrey, and Kent on its right bank. From its source in the Cotswold Hills to the Nore, the direct length is 120 miles, but with the windings it is probably 220 miles in length. In its upper course it passes through some of the finest agricultural country, while below London Bridge it is one of the most important commercial highways in the world.

The Thames has a very winding course, and between its source and Teddington there are numerous locks, or water-gates. A lock is an arrangement of two parallel

flood-gates, by which accommodation is secured between two reaches of different levels. One of the gates opens on the up-stream side, and the other on the down-stream side. The lock shuts in the water, and when a boat wants to pass through, one gate only is opened, and the boat is let in. Then this gate is shut, and the other gate is opened when the level of the water in the lock is the same as that of the water on the side where the boat wants to go. The boat is thus able to pass on to the next lock, when the same process is repeated. These locks are of so much importance in rendering the Thames navigable, that it is well to understand clearly their working.

The Thames first reaches Surrey at Runnymede, which is only a short distance from Old Windsor. Runnymede is famous as the place where the barons assembled in 1215, and Magna Charta Island is the spot where King John is said to have signed the Great Charter. There is a small cottage which contains a large stone on which it is said the parchment rested for the king and barons to place their signatures. It contains this inscription: "Be it remembered that on this Island in June, 1215, King John of England signed the Magna Charta; and in the year 1834 this building was erected in commemoration of this great event."

A short distance below Magna Charta Island is Bell Weir Lock, and at this point the River Colne falls into the Thames from its left bank. An immense quantity of water is here taken from the river for the water-supply of London, and a great reservoir has been constructed for storage purposes. Egham is the next town on the Surrey side, and opposite is Staines in Middlesex. This latter town is noteworthy from the fact that the London Stone, from which the town is supposed to have derived its name, marks the ancient boundary of the City of London.

Passing along the river we soon reach Chertsey, a place of some antiquity. There are remains of an abbey that was once the wealthiest in the kingdom. This is one of the few places where the Curfew Bell has survived, together with the old custom of tolling the day of the month after it has rung the close of day. The river now winds along to Weybridge, where the river Wey joins it. Weybridge once had a palace that was built by Henry VIII. Some remains may still be seen, and part of the park is now a garden.

Not far below Weybridge we come to Cowey Stakes, which is supposed to have been the scene of a battle when Julius Caesar crossed the river, 54 B.C. Some years ago, a number of wooden stakes were taken from the river. They were black with age, but quite sound. St George's Mill, quite near to the river, is the site of an ancient encampment. From its summit, beautiful views can be obtained extending over five counties.

Walton-on-Thames is an interesting place on the Surrey side, and the river is spanned by a modern iron bridge. Lower down the river is East Molesey, which takes its name from the river Mole, which enters the Thames at this place. On the opposite side of the river is Hampton Court, with its fine palace and gardens. Bushey Park, with its noble avenues of horse-chestnut trees, adjourns Hampton Court.

Thames Ditton and Long Ditton, two pretty Surrey villages, are passed as the river flows on to Surbiton, a town with many good residences along the river. Kingston-on-Thames is the next town of importance on the Surrey side. This town has a history reaching back to Saxon times. Some of our early English kings were crowned at this royal borough, and in the market-place is the "King's Stone" on which they are supposed to have been seated at their coronation. The river is here spanned by a beautiful stone bridge, which took the place of an ancient wooden structure. In the season, Kingston

Kingston Coronation Stone

forms the starting place for pleasure steamers up the river to Henley and Oxford.

Teddington Lock is two miles from Kingston Bridge and has been lately rebuilt. There is also a small lock for pleasure boats and a special one for barges.

8. The River Thames. (*b*) From Teddington to Deptford.

We have seen in the previous chapter that the Thames is navigable by boats from its mouth to Lechlade, and a proper depth of water is secured in the upper part of the river, as far as Teddington, by a number of locks which dam up the stream. If this were not done, the river would be swift and shallow, and not suitable for the passage of boats.

The lock at Teddington is the nearest to the mouth of the river; and above Teddington, which is about eighteen miles from London Bridge, the tide is no longer felt. The scenery on either side of the Thames is here very picturesque, and there are numerous river-side houses and beautiful gardens which render this part of the river extremely attractive. Twickenham has long been famous for its beautiful villas, and it was here that Pope lived and wrote and Horace Walpole had (at Strawberry Hill) his celebrated private printing press. Beyond Twickenham the Thames passes Richmond, one of the best known towns on its banks. The view from Richmond Hill is very fine, for it extends over a wide expanse of typical English scenery.

Poets have often sung the praises of the Thames at Richmond, and perhaps no English river has had so much written in its favour as the Royal Thames. Denham moralised on its beauties in the oft-quoted lines:—

> "Oh! could I flow like thee, and make thy stream
> My great example, as it is my theme!
> Though deep yet clear, though gentle yet not dull;
> Strong without rage, without o'erflowing full."

The Thames from Richmond Hill

There will be much to say about Richmond in another part of this book, so here we will deal only with the river. At Richmond a lock, known as a "half-tide" lock, has been recently built. Across the river there is a weir, whose gates are open only when the tide has been rising some hours, or during the first half of the ebb, when the water is still high. This New Tidal Lock keeps a sufficient quantity of water in the river between Teddington and Richmond, for before it was constructed in 1894, this part of the river was nearly dry at low tide. The weir at Teddington prevented the water coming down from above, and the ebbing tide took away the water below. However, this state of things has been altered since the New Tidal Lock was constructed.

It is about 68½ miles from Teddington to the Nore, and throughout the whole of that distance, the tide ebbs and flows four times in every twenty-four hours. The force of the tide is very great and its power can be seen at Blackfriars Bridge, where the water swirls round the piers, and rushes through the arches like a mill-race.

Kew is three miles beyond Richmond Bridge, and is world-famous for its botanic gardens, known officially as "The Royal Gardens." The river is here spanned by a new stone bridge, called the King Edward VII Bridge, for it was opened by our King in 1903. This part of the Thames is much frequented in the summer, both by boats and pleasure-steamers. Mortlake is the winning-post of the race which is rowed every year between the Oxford and Cambridge crews. The Boat Race has been rowed since 1839, and is looked upon as one of our national institutions.

Putney, the starting-point of the Boat Race, is also the headquarters of many rowing clubs and presents a scene of great activity in the season. It is connected with Fulham by a new bridge which was opened in 1886, and replaced an ugly wooden structure of 1729. The Thames now becomes a river of traffic and is spanned by many bridges, the last across the river being the Tower Bridge. From Putney to the Tower Bridge, the Thames flows through the busiest and greatest city of the world. The river has always played an important part in the life of London; and although it is not now used for business or pleasure as it used to be, it yet remains one of the chief arteries of the metropolis.

On the Surrey side, the Thames flows by Wandsworth, Battersea, Lambeth, Southwark, and Deptford on its way to the sea. Each of these places is densely populated, and although not in the modern administrative county of Surrey, yet they belong to it by history. Battersea with its beautiful park; Lambeth with its Archbishop's Palace; Southwark with its Cathedral; and Deptford with its Cattle Market, may all be seen from the Thames. These districts have undergone many and great changes, some of which we shall have occasion to mention later; but here we may refer to them as hives of industry, with crowded streets and great warehouses. The boundary between Kent and Surrey passes through Deptford, which was once the home of Evelyn and the place of residence of Peter the Great in 1698.

Before we leave the Thames, it is worth noting that the height of the tide in the river at London Bridge varies from 13 feet at neap tides to 21 feet at spring tides. The average duration of the ebb-tide is seven hours, and of the flood-tide five hours. It will thus be seen that the speed or current of the flood-tide is much stronger and more rapid than that of the ebb-tide.

9. Geology and Soil.

By Geology we mean the study of the rocks, and we must at the outset explain that the term *rock* is used by the geologist without any reference to the hardness or compactness of the material to which the name is applied; thus he speaks of loose sand as a rock equally with a hard substance like granite.

Rocks are of two kinds, (1) those laid down mostly under water, (2) those due to the action of fire.

The first kind may be compared to sheets of paper one over the other. These sheets are called *beds*, and such beds are usually formed of sand (often containing pebbles), mud or clay, and limestone or mixtures of these materials. They are laid down as flat or nearly flat sheets, but may afterwards be tilted as the result of movement of the earth's crust, just as you may tilt sheets of paper, folding them into arches and troughs, by pressing them at either end. Again, we may find the tops of the folds so produced washed away as the result of the wearing action of rivers, glaciers and sea-waves upon them, as you might cut off the tops of the folds of the paper with a pair of shears. This has happened with the ancient beds forming parts of the earth's crust, and we therefore often find them tilted, with the upper parts removed.

The other kinds of rocks are known as igneous rocks, which have been molten under the action of heat and become solid on cooling. When in the molten state they have been poured out at the surface as the lava of volcanoes, or have been forced into other rocks and cooled in the cracks and other places of weakness. Much material is also thrown out of volcanoes as volcanic ash and dust, and is piled up on the sides of the volcano. Such ashy material may be arranged in beds, so that it partakes to some extent of the qualities of the two great rock groups.

The production of beds is of great importance to geologists, for by means of these beds we can classify the rocks according to age. If we take two sheets of paper, and lay one on the top of the other on a table, the upper one has been laid down after the other. Similarly with two beds, the upper is also the newer, and the newer will remain on the top after earth-movements, save in very exceptional cases which need not be regarded by us here, and for general purposes we may regard any bed or set of beds resting on any other in our own country as being the newer bed or set.

The movements which affect beds may occur at different times. One set of beds may be laid down flat, then thrown into folds by movement, the tops of the beds worn off, and another set of beds laid down upon the worn surface of the older beds, the edges of which will abut against the oldest of the new set of flatly deposited beds, which latter may in turn undergo disturbance and renewal of their upper portions.

Again, after the formation of the beds many changes may occur in them. They may become hardened, pebble-beds being changed into conglomerates, sands into sandstones, muds and clays into mudstones and shales, soft deposits of lime into limestone, and loose volcanic ashes into exceedingly hard rocks. They may also become cracked, and the cracks are often very regular, running in two directions at right angles one to the other. Such cracks are known as *joints*, and the joints are very important in affecting the physical geography of a district. Then, as the result of great pressure applied sideways, the rocks may be so changed that they can be split into thin slabs, which usually, though not necessarily, split along planes standing at high angles to the horizontal. Rocks affected in this way are known as *slates*.

If we could flatten out all the beds of England, and arrange them one over the other and bore a shaft through them, we should see them on the sides of the shaft, the newest appearing at the top and the oldest at the bottom, as shown in the figure. Such a shaft would have a depth of between 10,000 and 20,000 feet. The strata beds are divided into three great groups called Primary or Palaeozoic, Secondary or Mesozoic, and Tertiary or Cainozoic, and below the Primary rocks are the oldest rocks of Britain, which form as it were the foundation stones on which the other rocks rest. These may be spoken of as the

Precambrian rocks. The three great groups are divided into minor divisions known as systems. The names of these systems are arranged in order in the figure with a very rough indication of their relative importance, though the divisions above the Eocene are made too thick, as otherwise they would hardly show in the figure. On the right hand side, the general characters of the rocks of each system are stated.

With these preliminary remarks we may now proceed to a brief account of the geology of the county.

At the outset, it may be said that the strata that are visible in Surrey belong to the later Secondary, or Mesozoic and the early Tertiary, or Cainozoic. There are old rocks, but these exist at such a great depth that they have no influence on the present landscape.

These very ancient rocks are only known to exist by boring operations; and at Richmond, Streatham, and other places where borings have been made, it has disclosed the fact that Devonian rocks are found at a depth of over 1000 feet.

Now leaving out of account these very ancient rocks, we shall find that the geological structure of Surrey is fairly simple, and easy to understand. The geological divisions of the county run generally from east to west in varying widths, and the old formations occur in the south and the newer in the north. This arrangement is due to the northerly dip or inclination of beds, caused by an unequal uplift of the land in past time, by which the south of the county has been raised higher than the north.

The oldest visible strata in Surrey are the Hastings

Beds of the Mesozoic Group, which occupy only a few square miles in the south-east of the county, just where the boundaries of Sussex and Kent meet. In this district we find beds of sand, sandstones, and coloured clays. The Hastings Beds dip northwards and are succeeded by the Weald Clay.

The Weald Clay is a very extensive formation in Surrey, and varies in width from five to ten miles. It is of river, not marine origin, with a thickness of as much as 1000 or 1100 feet, and it consists of bluish clay. Ironstone ore was once dug and smelted in this district of the Weald Clay. This was more particularly the case in Kent and Sussex.

The Lower Greensand is above the Weald Clay and

consists of yellow sands with beds of red clay. The Lower Greensand is remarkable in Surrey as yielding fuller's earth, which is found at Nutfield, near Reigate, in deposits 40 feet thick.

The Gault follows the Lower Greensand and is remarkable for the many fossils it contains. The Gault consists of dark-blue, stiff clay or shale which is used for bricks and tiles, and may be seen to the north of Redhill. The Upper Greensand is a formation that follows the North Downs in varying width through the county, from Titsey, Dorking, and Guildford to Farnham. This district is famous for its orchards and hop-gardens.

The Chalk formation which succeeds is one of the most interesting and important in the county. It forms the North Downs and terminates in the Hog's Back. The district is one of rounded, steep-sided hills, with deep, dry, winding valleys. The scenery is generally beautiful, and the North Downs reach a height of 895 feet. Sutton, Epsom, Leatherhead, Guildford, and Farnham are some of the towns that are built on the chalk; and sections of the chalk may be seen in various quarries and in some of the railway cuttings. The value of the chalk cannot be over-estimated, for it yields an abundance of lime and is one of the water-bearing formations.

We now come to the Tertiary or Cainozoic group. Following the Chalk formation, we have the Lower London Tertiaries. These have various names, but the Thanet Sands and the Woolwich and Reading Beds are chiefly represented in Surrey. These beds have a thickness of about 100 feet, and are not famous for their fossils.

The most important of the Tertiary formations in the valley of the Thames now follows. This is known as the London Clay, because it underlies the metropolis. It consists of a mass of bluish-grey clay, which is brown at the surface, and contains nodules of clayey limestone. The London Clay occupies a large area in Surrey, south of the Thames, and extends westward from Sydenham, Kingston, Walton, Cobham, and Guildford, to Aldershot.

After the London Clay we find the Bagshot Beds, which occur in the north-west of the county, from Esher to Aldershot by Bagshot to Egham. The formation is named from Bagshot Heath, where they are much in evidence. The sand is generally light yellow, and the district is mostly dry and barren land covered with heaths and fir. This is specially the case at Chobham Ridges, Fox Hills, and Pirbright.

The only formations left to consider are those of the most recent character. They are found along the valley of the Thames, as well as the other river valleys, and consist of mud, silt, and gravel. In these areas are found the remains of extinct animals; and as far as the valley of the Thames is concerned, there is evidence to show that once its estuary was far more extensive than it is to-day, and may have stretched as far as Windsor.

The soil of Surrey is so varied that it may be said to range from the richest loam to the barren heath and moorland. Speaking generally, the county to the west of a line drawn from Weybridge to Haslemere is a region of heath and moorland, of sand and gravel; and to the east of that line there are fertile river valleys and the Downs

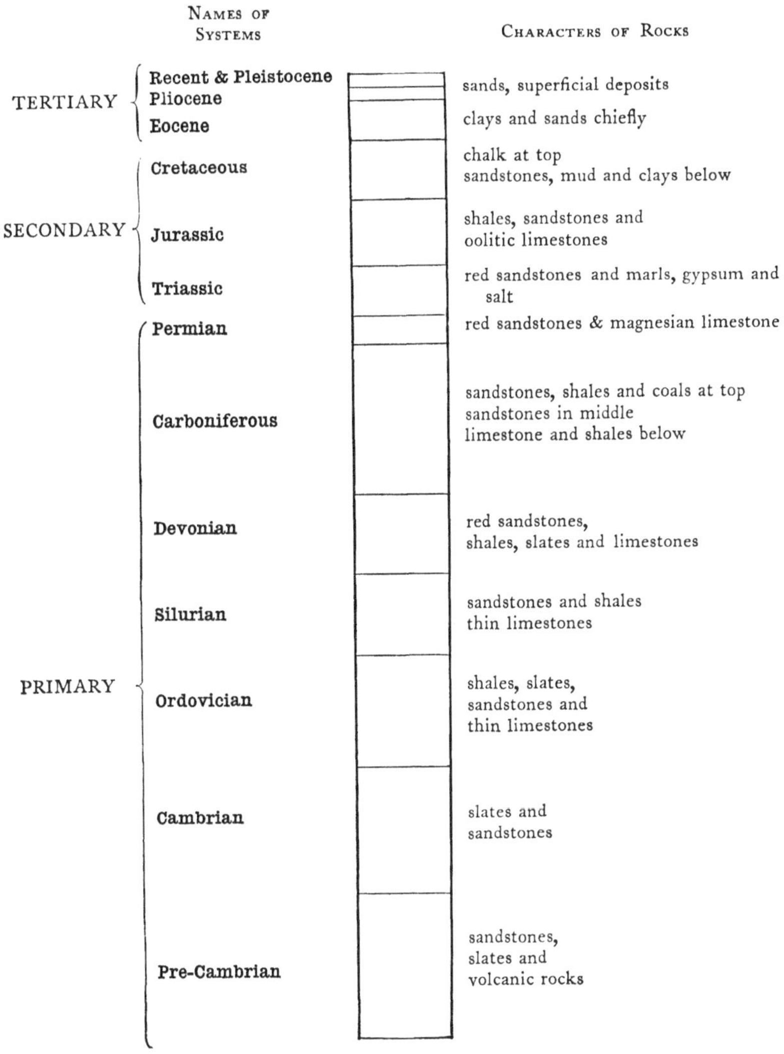

	Names of Systems	Characters of Rocks
TERTIARY	Recent & Pleistocene	sands, superficial deposits
	Pliocene	
	Eocene	clays and sands chiefly
SECONDARY	Cretaceous	chalk at top sandstones, mud and clays below
	Jurassic	shales, sandstones and oolitic limestones
	Triassic	red sandstones and marls, gypsum and salt
PRIMARY	Permian	red sandstones & magnesian limestone
	Carboniferous	sandstones, shales and coals at top sandstones in middle limestone and shales below
	Devonian	red sandstones, shales, slates and limestones
	Silurian	sandstones and shales thin limestones
	Ordovician	shales, slates, sandstones and thin limestones
	Cambrian	slates and sandstones
	Pre-Cambrian	sandstones, slates and volcanic rocks

with a thin soil and good pasture. To the south of the Downs there are the heavier clays and sands of the Weald, which yield good crops of wheat.

10. Natural History.

Various facts, which can only be shortly mentioned here, go to show that the British Isles have not existed as such, and separated from the Continent, for any great length of geological time. Around our coasts, for instance, are in several places remains of forests now sunk beneath the sea, and only to be seen at extreme low water. Between England and the Continent the sea is very shallow, but a little west of Ireland we soon come to very deep soundings. Great Britain and Ireland were thus originally part of the Continent, and are examples of what geologists call continental islands.

But we also have no less certain proof that at some anterior period they were almost entirely submerged. The fauna and flora thus being destroyed, the land would have to be restocked with animals and plants from the Continent when union again took place, the influx of course coming from the east and south. As however, it was not long before separation occurred, not all the continental species could establish themselves. We should thus expect to find that the parts in the neighbourhood of the Continent were richer in species and those farthest off poorest, and this proves to be the case both in plants and animals. While Britain has fewer species than France or Belgium, Ireland has still less than Britain.

The wealth of the flora of Surrey is a well-marked feature, and is mainly due to two causes, viz. the variety of geological formations, and the large area of waste-land, which covers about 114,000 acres, or one-third of the whole country. Perhaps the best way of illustrating the wealth of the flora is by stating the fact that out of 1861 native and naturalised plants in Great Britain, Surrey has no less than 1081.

Of the more interesting plants the box-tree deserves the first place. There is quite a forest of box-trees on Box Hill and there is no doubt that they are indigenous. The shrubs and forest trees are a most important feature in the botanical wealth of the county. Throughout the sandy soil of Surrey, Scotch firs and larches are very abundant, and many of the hills are crowned with these. Along the southern slope of the chalk downs yew trees are common, and in some of the churchyards, notably at Crowhurst, they have attained a great age.

Surrey is especially rich in brambles and ranks among the richest in Great Britain, Hereford and Devon being the only counties that surpass it in the number of species. Surrey has a large extent of bramble-ground, particularly the commons to the south-west of London, those near the Kentish border, and those in the south-west near the counties of Sussex and Hampshire.

The conditions that have favoured the variety of the flora of Surrey have also had a marked effect on the fauna, and it is probable that few counties of a similar area offer such an attractive field for the preservation of all

Kew Gardens; the Rockery

kinds of animals as Surrey. Reference will be made to some of the domestic animals in the chapter on agriculture, but it may be stated here that the wild animals of Surrey are similar to those that are found in most English counties. The red deer, once so abundant in Surrey, are still to be found in Richmond Park, where there are about 60 head of them. Fallow deer are still kept in many parts, notably at Richmond, Clandon, and Farnham. The roe deer remain only in the woods by Virginia Water.

The squirrel, dormouse, hare, and rabbit are very common in Surrey; and the fox, stoat, weasel, badger and otter are well represented. The badger is chiefly found, however, in West Surrey, and the otter in the rivers Wey and Mole. The wild cat, pine marten, and pole-cat, which lived down to recent times, are now quite extinct.

The fish of Surrey are essentially freshwater, but owing to part of the Thames being under tidal influence it possesses a few marine types, such as the herring, that extend into its limits. In the variety and abundance of British reptiles, Surrey stands second to no other county.

We cannot look in Surrey for any great abundance of the northern water-fowl or large birds of prey, such as we find on the firths and moors of Scotland or northern England. But taken as a whole the birds of Surrey are of very numerous species, and many of them of considerable interest. Without a seaboard, Surrey lacks the presence of many of the rarer maritime birds that frequent Kent and Sussex. It has, however, a number of lakes of fair size in the west and south-west, and to some extent it has connexion with the sea by the Thames; and so it affords a refuge for storm-driven stragglers from the coast, and a resting-place for migrants which would not otherwise visit this county.

FRENSHAM POND

Owing to game preservation and other causes, almost all the birds of prey have gone. The raven and the buzzard have disappeared, and the owl, the magpie, and the sparrow-hawk are decreasing. In many respects the black grouse was the most interesting bird of the county; but in the district round Leith Hill and Hindhead, where it was once so numerous, it is now almost, if not quite, extinct. Such charming residents as the kingfisher and great crested grebe are increasing, and speaking generally Surrey is a veritable paradise of small birds.

11. Climate and Rainfall.

The climate of a district depends on the temperature, the prevailing winds, the dryness or moisture of the air, and the character of the soil; and we may define the climate of a district as the state of that district with regard to weather throughout the year. Now in considering the climate of Surrey, we must bear in mind that it is an inland county, and so has not the modifying influences of the sea that are at work in Sussex and Kent. We must also remember that it is a small county, and so not subject to the varieties of weather that we should expect to find in so large a county as Yorkshire. Again, it is further south than Middlesex, and so, in the southern parts, the climate is generally warmer than the more northerly county.

It is of the greatest importance to have accurate information as to the prevailing winds, the temperature, and the rainfall of a district, for the climate of a county has considerable influence on its productions. Our knowledge of the weather is now much more accurate than it was formerly, and every day there appears in our newspapers a great deal of information on this subject. The Meteorological Society in London collects particulars from all over the country relating to the temperature of the air, the hours of sunshine, the rainfall, and the direction of the winds. Not only this, but it also issues weather forecasts, which are often correct. All these particulars are presented to us every day through the agency of the press, and some newspapers print maps and charts to convey the weather intelligence in a graphic manner.

There are also many stations to collect particulars of the rainfall, and these results are published in a book called *British Rainfall*, in which we find exactly stated the number of inches of rain that fell at certain stations. In Surrey alone there must be two or three hundred persons who keep a rain-gauge and enter in a register the daily rainfall.

We have, therefore, certain definite information about the climate of Surrey; and we shall find that even in this small county there is some variety in the climate. Thus we find that the east is drier than the west, and the north is colder than the south. The sandy districts have greater extremes of temperature than the heavy clay soil; and the presence of vegetation always tends to make the climate more equable. Sometimes there is a difference in the climate of two places that are not far apart. For instance one place may be on a hill-slope facing northwards, and another place on a hill-slope facing southwards. The latter place will of course be warmer, because it will receive the rays of the sun more directly than the former place.

Now let us take some of the recorded facts relating to Surrey temperature. In 1905, the mean temperature of England was 48·7°, and that of Surrey was 49·5°. Thus Surrey temperature is above that of England. If we look at the Epsom records of temperature, we find that the highest shade temperature in 1905 was in July and registered 86·2°, while the lowest, 17°, was in January.

Thus we get a range of 70° in the temperature of one place in 1905.

Next it will be interesting to study the amount of sunshine that gladdened Surrey and compare it with that of England. The average for all England was 1535·5 hours, while Surrey had 1667·2 hours of bright sunshine. Here again Surrey is in advance of England as a whole, but even in Surrey itself the number of hours varied, and we are not surprised to know that Haslemere had more sunshine than places in the valley of the Thames or near London.

The rainfall of England and Wales decreases as we travel from west to east. In 1906, the highest rainfall occurred at Glaslyn, near Snowdon, where no less than 205·3 inches were registered. The lowest rainfall in the same year was at Boyton in Suffolk, with a rainfall of 19·11 inches. These are both extreme records, but are useful for our purpose. Haslemere, in south-west Surrey, had 184 rain-days in 1906, and a rainfall of 37·53 inches. This was the highest rainfall in Surrey. A rain-day is one on which 0·005 inch or more is measured. At Guildford, which is further east, there was a rainfall of 25·36 inches on 146 rain-days; and at Richmond, which is in the north-east, rain fell on 150 days and measured 21·76 inches. The lowest rainfall in Surrey was at Esher, and amounted to 20·34 inches.

These records vary from year to year, but as they have been collected for a number of years, the average may easily be worked out for any place that has its station for the collection of weather and rainfall statistics. Thus

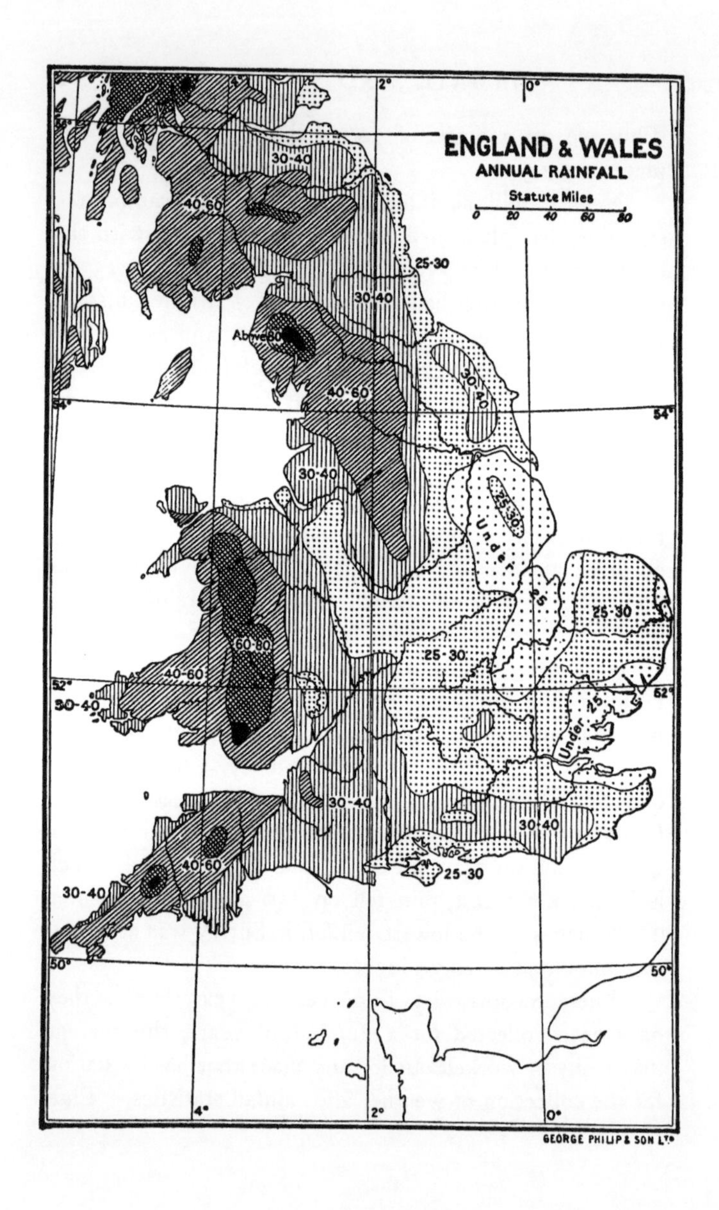
ENGLAND & WALES
ANNUAL RAINFALL
Statute Miles
0
20
40
60
80
30-40
40-60
25-30
30-40
Above 80
40-60
30-40
30-40
25-30
Under
25
25-30
25-30
60-80
40-60
30-40
Under 25
30-40
30-40
40-60
30-40
25-30
2°
0°
54°
52°
50°
4°
GEORGE PHILIP & SON LTD

at Kew the mean annual temperature for 24 years is 49·6°. For the same period the mean monthly temperature for January was 38° and for July 62°.

Now, to summarise the main facts in the climate of Surrey, we may say that the temperature of Surrey is warmer in summer and colder in winter than that of Kent and Sussex; and that the rainfall is heavier in the south and south-west than in the east and north. The climate in the Thames valley is damp and relaxing, and is rendered unpleasant by fogs in the autumn and winter. On the western highlands, the climate is bracing and healthy, and Hindhead and Haslemere are fine health resorts. In the Weald, which is sheltered from the cold north winds by the North Downs, the climate is generally bright and pleasant.

12. People, Race, Settlements, Population.

It is very difficult to give an accurate account of the earliest dwellers in Surrey. At first they were probably savages and quite ignorant of the use of metals. As time went on, they grew in knowledge and worked in stone, in bronze, and in iron. When Julius Caesar landed in Britain in 55 B.C. he found the Britons belonging to various stages of civilisaticn, of various races, and using different languages.

The most civilised people were in the south-east of Britain, and there is no doubt that some of the tribes had come over from Gaul. We shall not be far wrong in saying that the people in Surrey belonged partly to the Belgic Gauls, and that some of the older inhabitants of Britain were found in the forests, or places removed from the influence of the invaders. One of the tribes of the Britons was known as the Atrebates, and they lived in Hampshire, Berkshire, and West Surrey. Perhaps the Cantii, or people of Kent, lived also in East Surrey, for there was no natural barrier between the two counties on the east.

The Britons were subdued by the Romans, who held sway in our land for 400 years. When the Romans left in 410, Saxons, Angles, and Jutes came and settled in various parts of England. Sometimes these English came in a large body under the command of a great chief. They would conquer a district and then settle in its neighbourhood as one tribe. Thus we know the Jutes settled in Kent, the South Saxons in Sussex, and the Suthregienses settled in part of Surrey. They were not numerous, and came under the rule of the East Saxons and the Kings of Kent. Perhaps in Surrey, west of the Wey, the West Saxons had the rule, and it was not till much later that what we now call Surrey was constituted as a county.

Surrey is mentioned in the Domesday Book, and it is evident from the names of the parishes how strong had been the influence of the English conquerors. The place-names ending in *hurst*, *wood*, *holt*, *den*, and *fold* all bear testimony to the thoroughness of the English conquest. Ever since the time of the conquest Surrey has come under the influence of the metropolis, and of late years it has formed a kind of residential suburb for her merchants and clerks. We are not surprised, therefore, to find that the people of Surrey have not retained the same individuality that is met with in counties more remote. Here and there, in the south and south-west, some of the older inhabitants show traces of their early origin, and retain some of the peculiarities of an older dialect in their speech.

In common with other counties in the east and south-east of England, the social life of Surrey and the character of its industries have been affected by the immigration of foreigners. Flemings from Holland and Belgium, and Huguenots from France, took refuge in Surrey from religious persecutors in their own lands. They taught the people among whom they settled the art of weaving cloth, silk, and other textiles, besides introducing other industries of great value. It was chiefly in Southwark in the sixteenth century that the Huguenots settled, but they also formed colonies at Lambeth, Wandsworth, and Mortlake.

The influx of these skilled artisans from the Low Countries and France had the most favourable influence on the people of Surrey, who were taught industries that proved of the greatest value from every point of view. Their influence, too, on the character of our people was all in the right direction, for these foreigners were not only industrious, but thrifty and God-fearing.

From the foregoing account of the people of Surrey, it will be gathered that they are of mixed origin and race;

and this leads us to a consideration of the people of Surrey as we find them to-day.

When we come to the nineteenth century we are more certain of our facts with regard to the population of the county. It was in 1801 that the first census of our country was taken; and every ten years, since that date, there has been an enumeration of the population of each county, showing their age, occupation, and other interesting particulars. When the first census was taken, Surrey was larger in area than it is to-day. Then it included an area that is now embraced by the county of London. The modern administrative county of London was constituted by the London Government Act of 1899, and includes the civil parish of Penge and about 20 acres in Wandsworth.

The population of this administrative area was 653,549 in 1901, and shows an increase of 25 per cent. since 1891. The population of the entire ancient county, including the metropolitan parishes, was 2,012,744 in the same year, and it had risen to this figure from 268,233 in 1801. The number of persons to the square mile in the administrative county was 936, and in the entire ancient county 2655, against 558 for the whole of England and Wales.

The increase of population has taken place chiefly in the neighbourhood of London, and in such towns as Croydon, Richmond, and Wimbledon. The census shows that 464,258 persons were living in urban districts, and 189,291 in rural districts. In Surrey the females exceed the males by 47,023. It is worth noting that the people

Old view of Richmond Palace (now destroyed)

of Surrey live in 135,493 tenements, or houses, of which 69,787 have five or more rooms, and 35,706 have less than five rooms.

The census figures are very interesting, for they help us to form a vivid idea of the state of society at a particular period. Thus in 1901, there were 33,024 persons in Surrey whose age was 65 or over; and there were 8429 persons living in workhouses. Of the entire population, only 298,837 were born in Surrey; more than 114,000 were born in London: the bulk of the remainder were born in the other counties of England and Wales, Scotland and Ireland, and in the British Colonies. There were 6619 persons of foreign birth, and of these the greater number were natives of Germany, the United States, Italy, and France.

We also learn from the census returns a great deal about the occupations of the people. In Surrey, the men were chiefly engaged in house-building, in agriculture, as coachmen and servants, in the army, on the railways, or as clerks; the women were domestic servants, dress-makers, milliners, and teachers.

In bringing this chapter to a close, we would note that there are no convicts in Surrey, for the prison at Woking has been closed. The census returns reveal the fact that in 1901 there were 578 blind people in the county, and 256 who were deaf and dumb.

13. Agriculture. Main Cultivations. Woodlands. Stock.

We cannot consider Surrey as one of the agricultural counties like Essex, or Sussex, or Kent. It is rather a residential county, many of the people depending on the proximity of London for their occupation. Before the close of the eighteenth century, however, Surrey had some claims to be called an agricultural county, for it had its great land-owners depending upon farming for their rents, and there were many small farmers. It is generally stated that turnips were first extensively cultivated in this county. Arthur Young, who was an authority on agriculture, gives us some interesting particulars of the rotation of crops, the ploughing of the land, and the price of labour in the eighteenth century.

He found that crops were grown in the following rotation :—fallow, wheat, spring corn, clover, wheat, and beans, peas, or oats. Although he objected to this rotation, he found that 24 bushels of wheat were grown on an acre. The ploughing was done by teams of four horses, or of four, six, or eight oxen. Although ploughing-oxen may be seen in Sussex, they have quite disappeared from Surrey. Land was let at from 10*s.* to 15*s.* an acre, and wages of labourers were 1s. 2*d.* a day in winter, and 2*s.* and 2*s.* 6*d.* in harvest.

The farm labourer in Surrey in Young's day had usually a very bad house constructed of wattles and mud, though good timber and weather-tiles were not unknown. The labourer did not often eat wheaten bread, but was contented with barley and oatmeal. From the commons and heaths he got turf and brushwood for fuel, and on them he fed his geese. The common fields were then very extensive, especially along the northern edge of the North Downs, but towards the end of the eighteenth century Acts of Parliament were passed for their enclosure. Although many wastes and commons have thus been enclosed, much open land still remains, and the commons and heaths of Surrey add much to its charm as a residential county.

Having taken a glance at the condition of agriculture in Surrey in the eighteenth century, we can consider its position at the present time. Every year the Board of Agriculture issues a report on the acreage and produce of crops, and the number of live stock in each county. We will take that report for our guide and find what it has to tell us with regard to Surrey.

In 1905, there were 264,992 acres, or more than half the county, under crops and grass. The corn crops were wheat, barley, oats, rye, beans, and peas, which were cultivated on 48,569 acres, or nearly one-tenth of the county. Oats and wheat were the most important, and together account for more than 39,000 acres, while beans covered only 757 acres.

The green crops comprise, among others, potatoes, turnips and swedes, mangold, cabbage, and vetches, or tares. All these grow on 29,674 acres, and the first three occupy two-thirds of this acreage. About one twenty-third of the area of the county is devoted to the growth of clover, sanfoin, and grasses under rotation; and 157,985 acres are devoted to permanent pasture. This is by far the largest acreage, being more than one-third of the county.

No flax is grown in Surrey, but hops and small fruit together occupy 2160 acres. The remainder of the land that does not produce any of these crops is classed as bare fallow, and of this there were 6189 acres in 1905.

The growing of fruit, vegetables, and flowers for the London market is increasing in importance. Their cultivation is carried on mainly in the north of Surrey, specially near Mortlake and Ham, where large fields are laid out in rhubarb, cabbages, and other vegetables. Apples, pears, and cherries do not flourish greatly in Surrey, but strawberries are grown with increasing success.

The medicinal herbs grown in Mitcham and its neighbourhood deserve a short notice. In the open fields there, we find cultivated in large quantities mint, lavender, pennyroyal, camomile, liquorice, rosemary, and marsh-mallow. The perfume from these flower-farms pervades the whole country for some distance round.

The greater portion of the hop plantations in Surrey is at Farnham. Kent is pre-eminently the hop county; and probably Surrey does not produce one-fiftieth of the hops of the whole of England, while Kent produces two-thirds. Farnham hops still maintain a high reputation, though they were once more highly prized and celebrated than at present. The hop growers of Farnham cultivate only one sort, the white bine, and are specially careful in the drying and packing of the hops. The sandy soil of the district is peculiarly favourable to the cultivation of the

Ham House

hop-plant, and Farnham hops always fetch a high price in the market.

Before we leave the agricultural products of Surrey, we must glance at the woodlands and their extent. By far the larger part of the great forest of the Weald has long since disappeared, but to-day there are 54,437 acres of woodland in Surrey. The oak was one of the chief trees that grew in the Weald, and it was largely used for tanning, for charcoal, and other purposes. In the south of Surrey, the oaks are still very fine and very abundant; and on the sands, farther north, there are great plantations of fir-trees. On the chalk, the beech is the finest tree; and on Box Hill, as the name shows, the box-tree is very abundant.

Lastly, in studying the report on the Agricultural Returns, we must consider the animals that are used for various purposes. The live stock of Surrey are classified as horses, cows and other cattle, sheep, and pigs. Of these, sheep are most numerous, numbering 59,631, and very small this number seems when we compare it with the total for Kent, which is nearly 900,000. Cattle number 44,346, horses 12,290, and pigs 21,858.

Cows are reared to supply milk for the Metropolis, and cattle are fattened for the London market. The sheep are fed on the downs, the heaths, and moors, and are mostly of the breed known as the Southdown. They are noted rather for the quality of the mutton than for the quantity of the wool.

Box Hill

14. Industries and Manufactures.

There has been a great change since the days of the Conqueror in the wealth and importance of different parts of the country. A very large part of our chief industries are carried on in the northern counties, and Lancashire and Yorkshire have great masses of busy workers. The wealth of our coal and iron mines has been the chief factor in enabling us to secure our present supremacy as a manufacturing nation. These northern counties, too, are the chief seats of the textile industries, which is owing to the abundance and nearness of coal and iron, and the plentiful water-power. Thus, the north of England is the workshop of the world, and London is the great commercial centre.

In Norman times, however, there was a very different state of affairs, for neither coal nor iron formed an important item in English industry or trade, and the weaving trade was but little developed. Tin and lead were the chief mineral wealth, and raw wool and hides were the principal articles of trade.

We shall find in this chapter that the same thing has happened in Surrey, only on a smaller scale. Industries that were once important have ceased to be carried on, and other industries have succeeded them. Perhaps we shall also be able to find the reason for this change, which will be still more interesting.

There was a time in the history of Surrey when it was one of the wool manufacturing counties, and Guildford and Godalming were centres of the wool trade. It will at once be seen that both these towns were in the neighbourhood of the chalk hills, where sheep were fed in abundance, and so there was a large supply of wool. Every one had to buy a sign with a wool-sack painted on it to hang at the door; and the arms of Godalming are

Guildford High Street and Town Hall

still a sack of wool. It is quite common in some of the Surrey villages to see inns with the sign of the Woolpack or the Golden Fleece, which brings to mind the time when the fleece brought much wealth to England.

The weavers of cloth in West Surrey carried on their industry in their cottages. The wool was washed in the streams, then it was spun by the farmer's family, and finally made into cloth by himself and his sons. The cloth was often dyed blue with woad, which was specially cultivated for this purpose. Guildford cloth was much valued by Venetian traders in the sixteenth century, and was exported to the Canary Islands.

Early in the seventeenth century the woollen trade of Surrey began to languish, and the weavers were in a distressed condition. The trade has long departed from Guildford and the neighbourhood, for it is now cheaper to take the raw wool to the North of England where the coal is found than to bring the coal to such districts as Surrey where the wool is produced.

Glass-making was another industry that once flourished in Surrey. The Romans made glass at Chiddingfold, and remains of their manufacture have been found. In Henry III's reign, Laurence the glass-maker lived there: and in Elizabeth's reign there were many glass-houses which had to be suppressed, because of the nuisance caused by the smoke and smell. Besides Chiddingfold, Alfold and Thursley had glass factories, and in Alfold churchyard some French glass-makers are buried. Probably they were Huguenots who fled to England for safety, when their lives were threatened in France. Glass was manufactured in the neighbourhood of Chiddingfold, because of its natural advantages. The sand, the fern, the charcoal, and the firestone for the furnaces were all to be obtained in its vicinity.

Charcoal was extensively used for fuel, and many people called "colliers" made their living as charcoal-burners. The charcoal was employed in smelting iron, and of this industry there will be given a description in another chapter. Here it is worth noting that the gunpowder mills in Surrey owed their existence to the making of charcoal, which is one of the three substances in the old black gunpowder. The first powder-mills in Surrey were opened in 1554 or 1555, and in 1590 there were mills at Chilworth, which have continued down to the present day. Before that date, we got all our gunpowder from Flanders; and it is worth recording that it was some foreign immigrants who taught the English the use of saltpetre in the manufacture of gunpowder.

The Huguenots and other aliens from Europe brought their skill and instructed the people among whom they settled how to manufacture many articles that had previously been imported. Thus the Dutch from Holland and Flanders brought over the recipe for the brewing of beer, and to-day this is one of the chief industries in Southwark. Among the other manufactures introduced in the seventeenth century were hat-making, leather-dressing, and silk goods.

In north-east Surrey there were streams and oak woods, and so we find that the leather industry flourished in Bermondsey, where it is still carried on. At Godalming the tanning of leather is one of the chief trades of the town, for there the oak is one of the chief trees in the Weald and there is a good supply of water.

Lambeth was a manufacturing centre in the seventeenth century, and foreign workmen taught our English people the art of making plate-glass, delft ware, and earthenware. The latter manufacture is still one of the

The Thames at Richmond Bridge

principal industries, and Doulton ware is celebrated all over our country.

Wandsworth became a busy little manufacturing town in 1573, when a colony of Huguenots introduced the hat manufactory; and it is said that Wandsworth was the only place whence the cardinals of Rome could obtain a supply of their hats.

Coming to later times, we find that calico-bleaching and printing were carried on along the little river Wandle; and at Richmond, Mitcham, Croydon, and Wandsworth there were numerous factories for this purpose. In 1805, it is recorded that there were forty different industries along the Wandle, and that in its course of ten miles there were at least 3000 busy workers.

Nowadays the chief factories are in Metropolitan Surrey, where beer, leather, candles, soap, biscuits and other articles are manufactured on a great scale. There are, however, parts of Surrey where village industries are yet carried on. At Haslemere there has been an attempt in recent years to revive the weaving industry. Many women are employed in weaving linen, cotton, and silk, and producing curtains, towels, and dress material. This effort has so far been successful that tapestry, pottery, and other handicrafts are being fostered in this beautiful neighbourhood. The pottery works are quite worthy of special note, for almost every kind of decorative earthenware, pottery, and tiles are produced; and the red-glaze ware is specially beautiful.

15. Minerals. Exhausted Mining Industries.

There was a time when Surrey had its Black Country and smelted the iron ore that was produced in the district of the Weald. That period has long since passed away, and now Surrey's claim to rank as a mining county is of little importance. However, there are some very useful products, such as stone, chalk, fuller's earth, and clay, which may be considered under the head of minerals, and to which we may give some attention, before we deal with the past history of the iron industry of the county.

Surrey is not rich in good building stone, but both the Bargate stone and Firestone are of some importance in the county. Bargate stone is a conglomerate of quartz grains and pebbles, held together by a strong calcareous cement. In colour it varies from golden yellow to deep brown, and forms on the whole an excellent building stone. It is spread widely through the district about Godalming, and probably takes its name from Bargate, a few miles off, just across the Sussex border. The keep of Guildford Castle is built of this stone, and so are some of the churches.

Another stone that is quarried in Surrey is Firestone, which probably gets its name from its supposed power to resist fire. When first taken from the quarry it is soft, but soon hardens on exposure to the air. It is a greyish-green limestone found along the southern face of the Downs from Reigate to Limpsfield, and on to Merstham.

Henry VII's Chapel at Westminster, and parts of Windsor Castle, are built of it ; but its use is now almost entirely confined to the making of hearths and furnaces, for which its heat-resisting properties render it very suitable.

Guildford Castle

Chalk is one of the characteristic mineral products of Surrey, and is dug at many places for the making of lime and cement. At the beginning of the nineteenth century these industries were carried on at Dorking, Box Hill,

Merstham, Sutton, and Carshalton : now they are mainly confined to Dorking, Betchworth, and Oxted.

Fuller's earth is found at Nutfield, where it has been dug for centuries. It is of two colours, dark slate or blue, and yellowish brown. The blue is used by manufacturers of fine cloth, and is sent chiefly to Yorkshire. The yellow is employed in the manufacture of coarse woollen goods, and is sent to the north of England, Scotland, and Wales. Its great value lies in its power of absorbing oil and grease from the woollen cloths. It is easily prepared for the market, by removing every appearance of rust. It is dried by the fullers and ground in a mill before being used. Its use is of great antiquity, and so highly was it prized that a law of Edward II forbade its export. From the pits at Nutfield and the neighbourhood as much as 6000 tons have been exported in one year. Its use in the woollen manufacture has declined of late years, and it is now largely employed for toilet purposes.

For the purposes of brick-making, clay is dug chiefly in the south-eastern portion of Surrey. At the beginning of the nineteenth century, bricks were largely made at Kennington, Walworth, Camberwell, and Battersea; but as these places are now built over, the industry connected with bricks has gone to other districts.

At the commencement of this chapter we referred to a period when Surrey formed part of the Black Country of south-eastern England; and it is interesting to note that in the seventeenth century, the Wealden ironworks were by far the most considerable in all England. The ironworks of Kent, Surrey, and Sussex were of great antiquity, and there is good reason to believe that iron was worked in parts of these counties before the Roman conquest of Britain.

The industry flourished in the Weald because there was an almost inexhaustible supply of timber which could be converted into charcoal for fuel. In the Middle Ages the industry was in a most flourishing condition; and in 1319, Surrey and Sussex were ordered to provide 3000 horseshoes and 29,000 nails for an expedition against the Scots. About the fifteenth and sixteenth centuries the ironworks increased in importance, largely owing to the use of cannon in war; and in 1543 the making of cannon had become a notable part of the industry in the Weald.

It was somewhere about this period that the inroads upon the forest became very great; and we find that many complaints were made of the wasteful consumption of wood for making charcoal for the blast-furnaces, and of poisoning the air with smoke. Then it was that the Weald was the Black Country of the south-east of England.

In 1574, we get a list of the ironmasters of Surrey, and of their forges and furnaces. From this list we find that forges and furnaces were in operation at Cranleigh, Ewood, Copthorne, Lingfield, Burstow, and Haslemere. It is evident that the Surrey works were in full activity in Elizabeth's reign; and in 1588, after the defeat of the Armada, a commissioner was appointed to visit the furnaces and iron-forges of Surrey and Sussex, and to ascertain "the number and kind of the pieces of cast-iron ordnance now ready in the works." The owners and foremen were ordered not to cast any more such pieces of iron ordnance until they shall receive express direction from the Council. This prohibition was evidently from a fear that cannon might be exported to our enemies abroad.

Later in Elizabeth's reign it was enacted that no new ironworks were to be erected except within the former limits, and these upon old sites only, unless the owner could supply fuel entirely from his own property. This points to a fear that the timber in the Weald was being used extravagantly, and that measures must be taken to regulate the industry.

Notwithstanding these prohibitions, the iron industry continued to flourish until the Civil War. Then it met with a crushing misfortune, for Sir William Waller disarmed the Royalists in the south-eastern counties, and destroyed the ironworks belonging to them.

We cannot say exactly when the iron industry ceased in Surrey. Aubrey perambulated the county between 1673 and 1692, and speaks of ironworks at "Pope's Hole," and Witley Park. The making of cannon lingered on in the county till towards the close of the eighteenth century, but in 1809 a writer speaks of the ironworks as extinct.

Many local names in Surrey indicate the existence of the industries connected with iron. The water-power was often obtained by damming up a stream, and the water was then used to turn water-wheels, which lifted up and let fall heavy hammers. Thus we find the Hammer Ponds at Thursley, and Forge Pond at Lingfield, which owe their name to this fact. Among other names of similar origin we get Abinger Hammer, Hammer Farm, and Burningfold Wood where charcoal was made.

There yet remains in some of the old houses a good deal of the work of Surrey forges and foundries. Fire-dogs for wood fires and ornamented iron fire-backs are among the most common. The work is sometimes very good, and even graceful. Monumental slabs of iron are found both in Sussex and Surrey; and in Crowhurst churchyard, in the latter county, there is a cast-iron tombstone.

16. History of Surrey.

Our knowledge of Britain is very scanty before 55 B.C. and is chiefly gained from an examination of barrows, stone circles, and other remains of the earliest inhabitants. When, however, Britain comes into the light of history, we find the great Roman general, Julius Caesar, landing in Kent, and in 54 B.C. marching through that county into Surrey, in order to find a passage, or ford, across the Thames. He wanted to reach the Britons on the north of the river, and it was impossible to cross the Thames lower than Walton-on-Thames, in Surrey. Opposite this town there used to be a ford, now called Halliford, and this is the place where Caesar crossed with his army.

Cassivellaunus, the British chief, had defended this

Crowhurst Church

ford by placing sharp stakes in the bed of the stream, but it is not quite clear from Caesar's account how this obstacle was overcome. Perhaps they were partly removed by the cavalry whom he sent first into the water. They were followed by the legionaries, who went so swiftly and with such a dash, that the enemy, unable to resist the combined onset of horsemen and foot soldiers, left their stations on the bank and took to flight. Bede, one of our early historians, mentions these stakes, named Cowey Stakes, in the bed of the river, and later writers give minute details of them. However, they have long since disappeared, and are interesting only in connexion with the crossing of the Thames by Caesar.

We need not follow the campaign of Caesar, as it does not further concern Surrey; and although the Romans ruled England for 400 years, we do not hear of any fighting in this county. There is every probability that Surrey enjoyed a long period of peace and prosperity under Roman rule, and its people were, no doubt, much troubled when the Roman legionaries were withdrawn.

The Romans were followed by the Angles, Saxons, and Jutes, who first settled in and then conquered our land. Surrey was not a clearly defined kingdom, like Essex or Kent, and it would seem that it belonged at one time to Kent or Essex, and at another to Wessex. Surrey was between these kingdoms, and so received the attention of first one and then the other. At length the contest for Surrey came to a head, and in 568 A.D. a great and decisive battle was fought at Wipandune, which is probably Wimbledon, and there Ethelbert, King of Kent, was beaten by Cealwin, King of Wessex. For some time to come Kent was crippled, and Wessex enlarged its already extensive kingdom by annexing Surrey.

When the Saxons had conquered our land, they were harassed by the Danes and Northmen. The ravages of these invaders were specially severe in the ninth century, for we find, in 851, that three hundred and fifty ships came

"Caesar's Well," Wimbledon Common

to the mouth of the Thames. Their crews took Canterbury and London by storm, but their success received a severe check. They had crossed the Thames into Surrey, and were met by Ethelwulf and his son Ethelbald leading the West Saxon army. The battle took place at Ockley, on the edge of the downs which overlook Sussex, and there the West Saxon king, in the words of the *English*

Chronicle, "made the greatest slaughter among the heathen army that we have heard of till this present day, and there gained the victory."

The Danes, however, were stubborn foes, and their ravages and plundering continued in Surrey, even into Alfred's reign. It is said that they defeated Ethelred, in 871, at Merton, and it is certain that they plundered Chertsey Abbey. Indeed, through the greater part of Alfred's reign they were working much mischief in Surrey, although our great English king defeated part of them at Farnham.

The Saxon conquest of Surrey was, however, complete, and the Danish invasions did not make much difference in the end. The Saxons gave names to the towns, the villages, the farms, and the little streams, and to-day we can trace their influence in all parts of Surrey.

William the Conqueror was ruthless to the Saxons, who opposed him, and it is evident from the Domesday Book that many of the Saxon leaders lost their lands in Surrey. Many of the manors were henceforth to be the property of the Conqueror's friends; and Richard of Tonbridge, and Odo, Bishop of Bayeux, were liberally rewarded in Surrey for their devotion to their king.

As we pass on through our English history, we can spare time to glance only at some of the chief events that occurred in Surrey. Certainly the signing of Magna Charta, in 1215, at Runnymede, was an episode of the greatest moment, and we shall not be wrong in saying that it is the greatest event that has taken place in Surrey. Directly after John had agreed to the terms of the Great

Charter, he proceeded to break it. The barons in disgust invited Louis, the Dauphin of France, to take the English throne. In Surrey, Louis left his mark, for he took

possession of the three principal castles, at Guildford, Farnham, and Reigate.

The Black Death in 1348 was a terrible visitation, and wrought such havoc that probably half of the people of Surrey perished. Indirectly the Black Death was the cause of the peasants' rising under Wat Tyler, in 1381. The Kentish leader passed into Surrey at the head of his followers, and took possession of Southwark. The houses were plundered, the prisons were opened, and at length the gates of London Bridge swung back, and allowed the rebels to enter London. Wat Tyler was slain, and his men from Essex, Surrey, and Kent dispersed to their homes.

Jack Cade's rebellion, in 1450, began in Kent, and spread into Surrey; and the same thing happened in 1554, when Sir Thomas Wyatt raised an insurrection against Queen Mary. Wyatt plundered the Bishop of Winchester's house at Southwark, and on his way to Kingston the rebels alarmed the people of Surrey. Wyatt passed over the river at Kingston, marched towards London, but was taken and beheaded.

We now turn to a more glorious event in our national history, and to the fine spirit shown by the men of Surrey when our county was threatened by the Spanish invasion of 1588. It is an oft-told tale, but one which is ever fresh, how the Spanish Armada was defeated by the daring of our English seamen. Everywhere the greatest courage and determination were shown, but in the southern counties especially do we find this fine spirit displayed. Surrey had a muster of 6000 fighting men, and of this body many were sent to Tilbury, in Essex, where it was necessary to mass a large army for the defence of London. Surrey seamen went from Southwark, Bermondsey, and Rotherhithe to man the English ships. Above all, we must remember that Lord Howard of Effingham, the English commander, was Lord Lieutenant of Surrey, so that we are quite sure that Surrey took a leading part in the defence of our shores and in the defeat of the enemy.

Beacons were placed on the highest points in each county, and directly the Armada was sighted it was arranged that they should be lighted to convey the intelligence all over our land. Macaulay, in his ballad of *The Armada*, gives a picturesque account of the lighting of the beacons. From Devonshire he carries us to the home counties :—

> "The sentinel on Whitehall gate looked forth into the night,
> And saw o'erhanging Richmond Hill the streak of blood-red light,
> Then bugle's note and cannon's roar the death-like silence broke,
> And with one start, and with one cry, the royal city woke."

Macaulay then describes the bustle and movement in London and the surrounding districts, and pictures how

> "from wild Blackheath the warlike errand went,
> And roused in many an ancient hall the gallant squires of Kent.
> Southward from Surrey's pleasant hills flew those bright couriers forth;
> High on bleak Hampstead's swarthy moor they started for the north."

In Surrey, beacon-fires blazed from Hindhead, Hascombe Hill, South Hill, and Wimbledon Common, and roused the men of Surrey to a determination to banish the invader from our shores. How the Armada was defeated is a matter of history, familiar to most people; but it is worthy of note that Lord Howard of Effingham died in Surrey, and was buried in Reigate church.

The Civil War between the Royalists and the Parliamentarians did not affect Surrey to the same extent as some other English counties. Perhaps Surrey was equally divided in its allegiance, or opposition, to the King. Farnham Castle was held by the Royalists, but eventually suffered much at the hands of Sir William Waller, one of Cromwell's generals. With the close of the Civil War, and the Restoration that followed in 1660, our survey of the history of Surrey must end, for from this period its annals tell only of peaceful progress.

17. Antiquities—Prehistoric, Roman, Saxon.

We derive our knowledge of the people who first dwelt in Surrey not so much from written records as from the antiquities that have been found in various parts of the county. Compared with Kent, or Essex, these remains are not numerous; there are, however, sufficient to help us to understand something of the earliest dwellers in the portion of England we now call Surrey.

It will be convenient to consider these antiquities as they belong to the Stone Age, the Bronze Age, and the Iron Age. These three periods cover a wide extent of time, and it is not necessary to divide them off by years, for we cannot say when one age finished, and the next

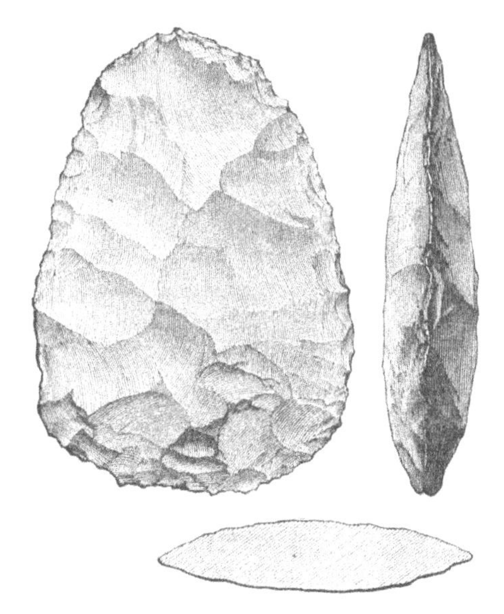

Palaeolithic Implement
(From Kents' Cavern)

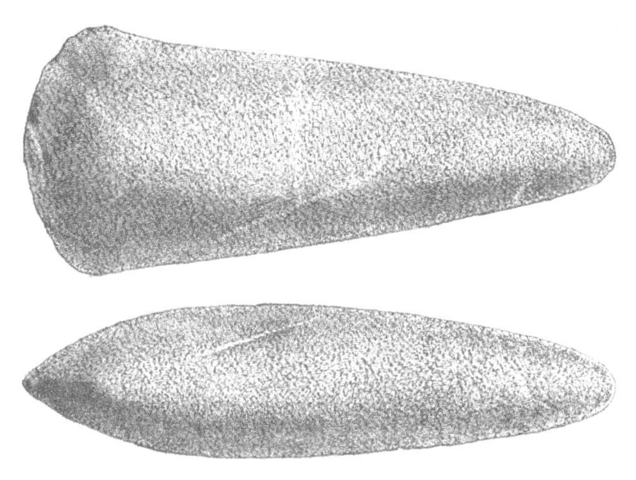

Neolithic Celt of Greenstone
(From Bridlington, Yorks.)

began. Following these three periods, we will consider the remains of the British, the Roman, and the Saxon, or Early English times.

The most ancient inhabitants of Surrey were the Stone Age men, who have left few traces except in the gravel deposits of the Thames valley, where flints rudely shaped are found with bones of the elephant, the elk, and other extinct animals. Of course, the North Downs supplied plenty of flints, which were fashioned into tools, weapons, etc. These remains of the Stone Age have been discovered at Guildford, along the Pilgrims' Way, at Croydon, Albury, and Chertsey.

As time passed on, these flints were more carefully worked, and arrow-heads, scrapers, flakes, and even polished celts were fashioned. By this time, early dwellers in Surrey had learnt the use of bronze and iron. Bronze celts have been discovered near Waverley and Crooksbury Hill, at Farley Heath, and Coombe Wood; and coins made of this metal have been found at Horley and Kew.

Probably the men of the Bronze Age and Iron Age were of a different race from those of the Stone Age, and we shall not be far wrong if we call them Britons, or Kelts. Of the early Britons in Surrey, we have remains in British camps, British barrows, and coins. British camps were places of refuge, where the women and children and the cattle could be kept while the men were fighting. They are generally on hill-tops and surrounded by banks and ditches. The best examples in Surrey are those at Anstilbury near Dorking, St George's Hill near Walton, on Holmbury Hill, and on Worm's

Heath. It is generally accepted that these camps all belong to the pre-Roman period.

The British Barrows are of two types—the long and the round, but no great numbers of either kind are found in Surrey, except near Addington, where twenty-five formerly existed near each other. On St Martha's Hill are three earth-circles, which may be relics of the worship of the Britons.

The Romans have left traces of their conquest in Surrey, and these may be classified under the following heads:—Roman villas and other buildings, Roman camps, Roman tiles, and Roman coins. There were no big Roman towns in Surrey, such as we find in Kent and Essex, but there is no doubt that Kingston was a Roman town; and a wooden bridge was thrown across the Thames at Staines, and another at Weybridge across the Wey.

Although Caesar marched through Surrey, he does not seem to have met with much opposition, nor do we read of any fighting with the Romans and Britons in this county. Hence the remains of Roman military rule are not numerous; but there is a rectangular camp on Puttenham Heath, called Hillbury, which has every appearance of Roman work. There is also a fortification on Farley Heath, which is considered to have been a Roman station.

There are remains of a few Roman houses between Reigate and Surrey, and at Southwark there were villas with tessellated pavements. Roman villas have been discovered at Bletchingley, Titsey, and Abinger, and at

Chiddingfold glass vessels have been found. Roman tiles have been worked into Guildford Castle and some of the older churches. The best and most enduring evidence of the Roman occupation are the fine roads which were made. Stone Street, which runs from Chichester to London, is in the best preservation of all the Roman highways in Surrey.

When the Romans left our land 410—436 A.D., they had done much to civilise the natives and teach them the arts of peace. Druidism had given place to Christianity, and when the new invaders came, the Britons were quite unprepared to meet them in warfare. The Angles, Saxons, and Jutes came first as pirates and ravaged the coasts, but eventually they settled in our land, and have left no uncertain traces of their conquest in Surrey. We have only to look at a map, or read over the names of the parishes, to recognise that nearly all the places owe their names to our Saxon forefathers. Let us take one good example, the affix *ham*, which means a home, and we can easily get a score of names with this ending, such as Ham, Bookham, Egham, Cobham, Mitcham, and Ockham. *Ley* is an English suffix denoting a meadow, and it occurs in Bramley, Frimley, and Bletchingley. Quite an interesting study it is to trace the origin of parish and village names, and it is most instructive to find that they are mainly of English derivation.

Some place-names in Surrey teach us that they were given to places where Saxon gods or goddesses were worshipped. Thus Tewsley was where *Tuis* was reverenced; and the god *Woden* may be traced in

OLD·MILL·AT
DORKING·

Woden Hill. Thursley may be connected with *Thor*, Thundersfield with *Thunor*, or *Thor*, and Sattersley with *Saturn*.

Besides these traces of the Saxon conquest, there are many other evidences of their settlement in Surrey. Thus there are earthen mounds and ditches that were constructed for defence; and there are burial places from which Saxon ornaments and weapons have been obtained. Saxon coins have been found at Winterfold Hanger, near Dorking, and swords, spears, and other weapons have been obtained from various parts of Surrey.

Kingston, or "King's Town," was of great distinction during the Saxon period, for we find that eight kings were crowned in it. In the open space in front of the Court-house there is placed the ancient stone upon which the Anglo-Saxon Kings were enthroned during their coronation. The King's stone, a shapeless block, is placed on an octagonal base, on the sides of which are inscribed the names of the eight Kings crowned on it. This historical relic is enclosed within an ornamental railing supported by stone shafts.

18. Architecture. (*a*) Ecclesiastical—Churches.

The ecclesiastical buildings of Surrey are not so numerous nor so important as we might expect. This is partly owing to the large tract of forest land which covered Surrey in the south, and to the extensive uncultivated areas in the west, which were unprovided with churches. Again, many of the churches were constructed largely of wood, and their details were not so finished as in such counties as Kent and Sussex. We shall, however, find much that is of deep interest in studying the ruins of the religious houses and the ancient churches that are still in use.

Chertsey Bridge

At the outset we must clearly understand that Surrey had no great ecclesiastical centre. Kent had Canterbury and Rochester, and Sussex had Chichester ; but it is only quite recently that a cathedral has been constituted at Southwark. Surrey has never formed a separate ecclesiastical diocese, but has mainly belonged to Winchester. Hence the influence of Winchester has been very great, especially in that part of Surrey that borders on Hampshire. In the south, some of the churches are similar in construction to those of Sussex, while on the north, between Croydon and Chertsey, there is great similarity between the Surrey churches and those of London and Middlesex.

In the construction of its buildings, both ecclesiastical and domestic, Surrey has suffered from a scarcity of really good building stone, and in the period before the Conquest, wood was extensively used. The roads were bad, and as there was an abundance of timber, we are not surprised to find that the builders of those days did not go to the expense of importing stone into the county. In the ancient churches of Surrey we find numerous timber towers, porches, and bell-cotes; and in Haslemere Church the pillars were of oak.

For the roof-coverings of the oldest churches thatch, tiles, stone, slate, and lead were employed, sometimes with remarkably good effect. In the greater number of the ancient churches flints were used in the walls; and in such churches as Esher, Leatherhead, and Mickleham there is some good flint and stone chequer work.

In south-east Surrey a good building stone, or "firestone" as it is called locally, is used in the construction of the churches; and in south-west Surrey, Bargate stone is employed with success. It is a hard sandstone, of good wearing qualities, and has been in use for hundreds of years. Very little Caen stone is found in the Surrey churches; and Sussex and Purbeck marbles are used very sparingly.

Churches built before the Reformation were rarely of brick construction, but after that period we have the churches at Kew, Kingston, and Petersham, mainly built of red bricks. Archbishop Whitgift's almshouses at

Whitgift's Almshouses at Croydon

Croydon, and Archbishop Abbot's almshouses at Guildford, are some of the finest and most picturesque of red-brick buildings in Surrey.

It is interesting to note that there are 145 churches in Surrey that were built before the Reformation, and of this number, sixty-four are mentioned in Domesday Book. Of these oldest churches that were built before the Conquest, perhaps the most noteworthy are at Ashstead, Fetcham, Godalming, Guildford, St Mary's, and Stoke D'Abernon, and although they have undergone much restoration, portions of the earliest period may yet be traced.

All the old churches of Surrey are marked by a general simplicity in plan, which is cruciform or a plain parallelogram; and the arrangement would be chancel, a central tower, and a nave. At Compton there is a very curious church with a double chancel. There are fifty churches in Surrey with a stone-built tower at the west end, and thirty churches have wooden towers or turrets.

Now that Southwark has recently become the see of a bishop, we may well include St Saviour's Cathedral in this chapter on the ancient ecclesiastical buildings of the county. Of course, Southwark is now part of the county of London, but its historic associations will always belong to Surrey.

St Saviour's Cathedral is opposite London Bridge Station, in the Borough High Street. The original Norman nave, of which fragments remain, was built in 1106, by Giffard, Bishop of Winchester, and was the Church of the Priory of St Mary Overy. In 1207, another Bishop of Winchester built the choir and altered the nave; and in 1540, Henry VIII converted it into a parish church. Since then, the church has been almost entirely rebuilt, and is now the Cathedral of the diocese of Southwark.

Heretics were tried in St Saviour's Church in Queen Mary's reign, and many interesting monuments remind us of the great men connected with it. John Gower, one of our earliest poets, was buried here; and Lancelot Andrewes, Bishop of Winchester, rests in the Lady Chapel. Massinger and Fletcher, and Edmund Shakespeare, brother of the great dramatist, are also buried here. These names recall the fact that the Globe Theatre was near the present cathedral, and that William Shakespeare had a house in Southwark.

19. Architecture. (*b*) Ecclesiastical—Religious Houses.

Our knowledge of the ecclesiastical architecture of Surrey would be very incomplete if we did not consider the religious houses that once existed in the county, but of which we can now see only the ruins. Before the Reformation England was dotted over with abbeys, monasteries, and other religious houses, which were often fine specimens of the architect's skill. In many parts of our land there yet remain enough of these buildings either entire, or in ruins, to impress upon us their beauty, and to convince us of the money that was lavished in their construction.

In many instances, the religious houses were centres of religious influence; they were the chief seats of learning, and they were the homes of men and women who helped the peasants, and promoted the prosperity of the countryside. Henry VIII with ruthless hand, and guided by such advisers as Thomas Cromwell, decided to close the lesser monasteries first, and the greater monasteries afterwards. Attached to the religious houses were hundreds, and in some cases thousands, of acres of the best land; and the rent from these lands, together with other riches, helped to make the abbeys and monasteries very wealthy. It was no doubt with a view to enrich himself and to reward his courtiers that Henry VIII laid violent hands on the religious houses of England. From his point of view, he was successful; but his harsh action told very heavily on the poor throughout the country, and it is now generally felt that Henry was not moved by a spirit of piety, but rather by a love of money, in thus suppressing the monasteries.

As far as Surrey was concerned, the number and value of the religious houses were not so great as in such counties as Kent; but we know enough of them to be certain that they represented a very important factor in the life of Surrey before the Reformation. We may be sure that they were civilising agencies of no mean order, and probably conveyed to the great mass of the people some of the best elements in social and religious life.

It is unfortunate that there are so few remains of these beautiful buildings in Surrey. There yet may be seen the ruins of three or four, but of at least a dozen others not a stone remains, and in some cases it is even difficult to ascertain the site of the building. With one exception, that of Sheen, all these religious houses in Surrey were built before the thirteenth century, so that they represent the earliest period of ecclesiastical architecture.

The most ancient of these buildings was Chertsey Abbey, which was founded by Erconwald, Bishop of the East Saxons, in 666. Thus Chertsey was established as a religious centre soon after Augustine had landed in Kent, and probably marks the beginning of the re-introduction

The Wey near Byfleet

of Christianity into Surrey. The Abbey at Chertsey was richly endowed with broad acres in north-west Surrey, and soon took its place among the great houses of the religious order in England.

It suffered much at the hands of the Danes, who pillaged and burnt it, and killed ninety of its monks. It was, however, rebuilt, and other monks continued the work which had been begun. The Thames was embanked, the Wey was bridged, mills were built, fish ponds were dug, cottages were built, and a vineyard was planted. All this was done under the direction of the abbots, and when the Abbey was dissolved by Henry VIII, its revenues were very large.

If Chertsey Abbey was the earliest religious house in Surrey, Waverley Abbey, near Farnham, was the most interesting, both historically and architecturally. Waverley Abbey was founded by Bishop Giffard, of Winchester, in 1128, and the first monks came from Normandy. They belonged to the order known as Cistercians, who had many beautiful houses all over England. Their buildings were generally in country places, by the side of streams, in wooded valleys. They formed places of rest for the wayfarer and the pilgrim, and the farms were always well cultivated.

King John visited this Abbey in 1208, and was royally entertained. Simon de Montfort was also received in 1245, and Edward III showed it great favour. When the magnificent church was finished in 1278, thousands of people were feasted for the space of nine days, at the end of which time, they "returned to their homes glorifying and praising God."

The end of Waverley came in 1536, when it fell with the lesser monasteries. The abbey and church fell into ruin, and the stone was carried away to build houses in the neighbourhood. Now there are a few ruins above ground, and the bases of the walls. All the old accounts picture Waverley Abbey as one of the stateliest buildings, but its glory has long since departed.

An interesting account of the ancient kitchen-garden of the monks is given by Cobbett. He says, "It was the spot where I first began to learn to work, or rather where I first began to eat fine fruit in a garden; and though I have now seen and observed upon as many fine gardens as any man in England, I have never seen a garden equal to that of Waverley."

We must now turn to another religious house known as the Priory of Merton, which was founded in 1114, and of which some remains may yet be seen. A window, a gateway, and part of the walls are the remnants of a building that was once very famous. It was at the College of Merton that Thomas a Becket was educated; and from it went forth Walter de Merton, the Chancellor of England, who founded Merton College, Oxford. Hubert de Burgh found refuge here, and in 1236 a Parliament was held at Merton.

Newark Priory has the most considerable ruins of all the Surrey religious houses. The Priory stood on the north side of the River Wey, not far from Pirford. The grey walls are very thick, being formed of flint bound together by very hard mortar. The principal ruin is probably a part of the Priory church, which was built at the end of the twelfth century.

Sheen Abbey was the last religious house to be founded by Henry V, but of this and the smaller religious houses at Guildford, Reigate, Sandon, and Lingfield, there is no need to write at length. The names survive, but the buildings have long since gone. Perhaps Bermondsey Abbey is worth mentioning, as its monks embanked the Thames to save Bermondsey and Rotherhithe from floods. The monks of Bermondsey and St Mary Overy founded St Thomas's Hospital.

Ruins of Newark Abbey near Ripley

20. Architecture. (*c*) Military—Castles.

Surrey cannot boast of great castles such as we find at Dover or Rochester in Kent, or at Pevensey in Sussex; but there is quite enough of interest relating to Surrey castles to make an instructive chapter. The building of castles is usually associated with the early Norman Kings, but castles or strongholds were built by the Saxons previous to the Norman conquest. Perhaps the very earliest idea of these strongholds may be traced in the camps that were formed on hills. Some church-towers were used as places of defence, and often the old houses were surrounded by moats. Thus we find at the present time several old manor houses that have moats, such as Tangley Manor at Wonersh, and Cadworth Manor near Newdigate.

It was during the Norman period, however, that castles were chiefly built, and it is said that not less than 1100 were raised during a period of about sixty years. These castles were frequently built on the site of Saxon strongholds, and were the fortresses of the barons who owned the neighbouring country. Too often the castles became the terror of the country-side, and men rejoiced when Henry II gave orders for many of them to be destroyed.

Most of the great castles were built on the same plan. First there was the donjon, or keep, on either side of which was the inner ballium, or bailey, and the outer ballium. The whole area was enclosed by high walls, with one or more gateways, which were entered by a drawbridge. Among other buildings there would be the stables and the chapel, and often in the outer ballium a mound was raised, which formed the court-hill or the place where justice was executed. Where possible, the castle was usually surrounded by a moat.

When we consider the position of the Surrey castles, we notice that the chief ones were built along the North Downs, and so formed a line of defence right across the county. Now look at a map and you will observe how well the sites were chosen. At the west of the county there was a castle at Farnham; another was raised at Guildford, by the side of the Wey; a third was built at Reigate, near the Mole; and right in the east was Bletchingley. Of all the Surrey castles, those at Guildford and Farnham were the most important, and of them there stand considerable ruins, to enable us to judge of their former greatness.

Let us take these castles in order, beginning with Farnham. This great castle was built by Henry de Blois, in 1136, and its situation was chosen because it commanded cross roads and because it had been fortified in earlier times. There was originally a great mound at Farnham, and it was on this that Henry de Blois raised his keep. In the reign of Henry III, Farnham Castle fell into the hands of Louis, and was afterwards dismantled. During the Civil War it suffered great damage, but has been subsequently restored. It is now the residence of the Bishop of Winchester, and the servants' hall is part of the original building.

Guildford Castle occupies a fine position, standing as it does on the chief line of communication running north and south through the gap in the Downs. Here, too, as at Farnham, there was a great mound, before the castle was built on it some time in the reign of Henry II. The Norman stone keep still stands, and, with the surrounding pleasure grounds, is the property of the Town Council.

Henry III was often at Guildford Castle, and like Farnham and Reigate Castles, it fell into the hands of Louis of France. It had no share in the Wars of the Roses, or in the Civil War, and there is no record that it ever stood a siege. For many years it was the place where the prisoners from Surrey and Sussex were confined; and from 1612 to 1885 it was in private hands.

Reigate is the third of the castles built as a defence on the line of the North Downs. Here there was no mound on which to build, and there is some uncertainty

Reigate Castle

as to whether the castle was built before or after the Conquest. Its position was naturally a strong one, for it commanded the route through the gap in the North Downs. It was captured by Louis of France and the English Barons in 1216, and was destroyed in the middle of the seventeenth century, when its stone was used to repair the roads. The ruins of the castle belong to the town, and the grounds are laid out as public gardens.

Bletchingley Castle is the last in the line of fortresses along the North Downs. Its date of erection is uncertain, but we find that it was injured in the thirteenth century. In the reign of Henry III this castle was much used, and no doubt its square Norman keep, two ditches, bank and wall, were the scene of many deeds of valour. Two hundred years ago there were remains of stone walls; but now nothing but earthworks and foundations remain.

Farnham, Guildford, Reigate, and Bletchingley were the greater castles of Surrey. They stood near the Pilgrims' Way, where this ancient road is cut by cross-roads from north to south. Besides these four castles, there were strongholds or fortresses at Kingston, Ockley, Addington, Thundersfield, Lingfield, and Betchworth. There is, however, little of these ancient buildings to be seen; sometimes only the moat or earthworks remain, while in other cases there is nothing but the site to remind us of the fact that once a castle ruled the neighbourhood.

21. Architecture. (*d*) Domestic—Famous Seats, Manor Houses, Cottages.

In the previous chapter we considered the castles and fortresses of Surrey, and found that they were built by the great nobles and barons as a protection for their possessions. In course of time, however, as the country

Surrey Cottages

became more settled and peaceful, the nobles ceased to live in their castles, and built themselves large houses, which were more comfortable as places of residence. Not many of these old houses now stand, but there are parts of England where buildings dating from the fifteenth and sixteenth centuries may yet be seen.

In many parts of Surrey there are houses and cottages that have distinct artistic merits, and a certain individuality that makes them pleasant to the eye. Nowadays we are accustomed to think of cottages and houses as we find them in the suburbs of London, or in some of our growing towns. There we find long straight streets, with rows of houses that have no distinctive individuality. They are built after one pattern, and their walls of yellow brick and roofs of dull slate are certainly not artistic.

It is a real pleasure to walk up the High Street at Guildford, where the houses and shops are of varied design; and as they are built of different materials, each one seems to reveal the character of the builder. Again, if we go into some of the Surrey villages, as Chiddingfold, Witley, or Haslemere, we find cottages which are entirely good for the eye to rest upon. These old Surrey cottages have grown beautiful with age, and being in the midst of delightful gardens, we at once realise how sadly our present day cottage architecture is behind that of two or three hundred years ago.

It may be said at once that Surrey has no great architectural buildings, like Hatfield House in Hertfordshire, or Knole House in Kent. There were famous palaces at Richmond and "Nonsuch," but these were destroyed long ago. Southwark, too, had palaces and great houses of famous men, but they have also passed away. Guildford and Godalming retain many historical buildings, and in the High Street of Guildford may be seen such fine specimens as Abbot's Hospital, whose bricks have now weathered to a generous red. In the region of the Weald, on the borders of Kent, Sussex, and Hampshire, small manor houses and farm-houses are to be found.

The architecture of a county is always influenced by the building materials that are found within its borders. We naturally expect to find in the Cotswold district, houses not only built of stone, but roofed with stone. Similarly, we expect to find the houses and cottages in Surrey largely built of timber, owing to the ease with which this material could be obtained from the Weald and other forests. Other materials were used in various parts of Surrey. For instance, in the eastern parts, Kentish rag or "firestone" of a strong greenish-grey colour is worked; while to the south of Guildford, Bargate stone is employed. Black ironstone is found in the sand districts, and in many parts there is plenty of chalk. Although bricks are now freely used, we must remember that after the Roman occupation they were not employed till about the early fifteenth century.

Perhaps it would be correct to say that the good houses of Surrey were often built of stone, while the offices and outbuildings were of post and panel. Smallfield Place at Reigate, and New Place at Lingfield, are excellent specimens of such houses. Lambeth Palace, the London residence of the Archbishop of Canterbury, is a fine example of fifteenth century brickwork, while the gate tower at Farnham was built about 1500.

We cannot take a better example of a timber-frame house than that of Tangley Manor. It was built round an open hall in 1582, and its ornamental front is entirely pleasing. It may be noted here that up to the middle of

Sutton Place

the fifteenth century, all such houses had a lofty hall running up to the roof. An outer door led into a passage cut off by a screen from the hall. From this passage, two or three doors opened to the offices, such as pantry, buttery, or cellar. The same plan of construction is still to be seen in most of the colleges at Oxford and Cambridge. In the sixteenth century, rooms were built over the hall, and a fireplace and chimney took the place of the central open hearth. Later on, other rooms were added, and better stairs led up to the top rooms.

One of the most interesting houses of Surrey is Sutton Place, near Guildford. It was built in the reign of Henry VIII, and originally formed a quadrangle. Sutton Place is of red brick, with mouldings of terra cotta. Some of the windows are beautiful in form and proportion, and the windows of the great hall, which extends throughout the centre, contain some shields of arms. Although the house has been modernised, there yet remains much of the older fittings, and it may be justly regarded as one of the fine mansions of Surrey.

Loseley, another Surrey mansion of great interest, was built about 1562. The house is of grey stone, large and stately, and is a good example of an Elizabethan mansion. Some of the rooms have very fine work, but the great hall is the chief attraction.

There is one feature in some of the old Surrey houses and in many of the country cottages that is of great interest. The brick chimneys are very varied, and in such a building as Abbot's Hospital at Guildford, they are fine and dignified. In cottages of wood construction and plaster filling-in they are nearly always on the outside walls and are of great bulk.

The bricks and tiles made in Surrey were of excellent character and colour, and we find in the south many picturesque brick buildings. Many of the cottages have

Loseley Hall

tiled roofs, but in parts the houses are covered with stone slabs. Some of the Surrey cottages are noted for their weather-tiles, with fancy ends, which are fitted to the outer walls.

THE ARBOUR SUTTON PLACE

In this chapter, we have been referring entirely to the Surrey houses and cottages that were built before the

eighteenth century. With regard to the modern houses, it is not necessary to make any reference, as they have not the charm or picturesqueness belonging to the older buildings.

22. Communications—Past and Present —Roads, Railways, Canals.

It is rather difficult to believe that less than two centuries ago London Bridge was the only bridge over the Thames from Surrey to Middlesex; that Putney Bridge, which was the next to be constructed after London Bridge, was not finished till 1729; and that Westminster Bridge was not opened till 1750. Yet such is the case, and we must try to realise the primitive character of the communications in Surrey only a century ago. Most of the roads were very bad; there were no railways; and the canals were of little importance. All that is changed, for Surrey has now some of the best high roads in England, an excellent system of railways, and many fine bridges over the Thames.

Let us go back to the earliest times and find what traces there are in Surrey of the old British trackways. Perhaps we shall not be wrong in starting with a road that led from Southampton through Winchester and so through Farnham, Guildford, and Dorking into Kent, to its destination at Canterbury and the Kentish sea-ports. This must have been the route taken by merchants who landed their goods at Southampton and conveyed them overland through Hampshire, Surrey, and Kent. Later, this route was followed by the pilgrims from the west and south-west of England who journeyed to Becket's shrine at Canterbury. The Pilgrims' Way, as it is well called, leads through a most interesting part of Surrey, and the important thing to notice about it is its position

Lane near Dorking

on the high, dry ground of the Downs, and that it avoids the large towns by which it passes. In its course through Surrey, the Pilgrims' Way is often bordered by yew trees, which attain a great size in the chalky soil.

Other British trackways have been traced at Wimbledon, and from the mouth of the Arun northwards to the Thames. There is no doubt that some of these British tracks were improved by the Romans, who were the great road-makers of our country. A glance at a map of Surrey will show that the chief roads run from south to north, and the reason for this is evident. The main roads of Surrey are between places outside the borders of the county, and it will be seen that the Romans carried their roads from openings in the sea-coast of Sussex, such as the mouth of the Ouse, the Arun, and Pevensey, through the gaps in the South Downs and the North Downs onwards to the Thames.

We have not space to trace all these Roman roads, but we will follow one, Stone Street, which ran from the mouth of the Arun. This was probably an old British trackway which they paved with stone, and so the street got its name. It passes Ockley Green and runs on quite straight to Dorking and Burford Bridge, where it is called Pebble Lane. Its course is now up the Downs to Epsom, and straight past Ewell and Streatham ("the Home on the Street") to London.

Other Roman roads entered Surrey at the north-west and north-east corner, coming from Hampshire and Kent respectively. The former came across Bagshot Heath to the Thames at Staines; and the latter came towards London, and is now known as the Old Kent Road.

Now we must leave the times of the Romans and come down to a more recent period. There is not much evidence as to the formation of new roads through Surrey till the nineteenth century. We do know, however, that the chief roads of the county were in a shocking condition in the seventeenth and eighteenth centuries. It is related that the people of Horsham wanted a new road to London in 1750, for they stated that to reach London from that town, it was necessary to go down to the sea-coast and reach London by way of Canterbury.

A little later, in 1755, the present road from Capel to Dorking was constructed, and tolls were levied on carriages drawn by six horses. This fact alone shows the condition of the roads, and we know, too, that oxen were used to draw carts and carriages through the Wealden clay. From 1755, a new era commenced, and many good roads were made in Surrey, and communications were generally improved. The well-known Brighton road was completed in 1807; and the equally famous Portsmouth road, now so much used by cyclists and motorists, was much improved in the last century.

Besides the scarcity of main roads in the early eighteenth century, we find, as late as 1780, that there was only one road between Epsom and Croydon, and none between Sutton, Chipstead, and Caterham. At the present time, the county is traversed by five great roads, viz.:—London to Brighton by Sutton and Reigate; London to Brighton by Croydon; London to Portsmouth by Ripley, Guildford, and Godalming; London to Guildford and Farnham by Epsom and Clandon; and the Salisbury and Exeter Road by Staines and Bagshot.

We will now turn aside from the earliest communications by means of roads, and devote some attention to the canals and railways of the county.

It is generally agreed that Surrey was the first county

to be provided with canals. The navigation of the Wey is artificial, and the river has locks which are supposed to have been the first constructed in England. It will be noticed that the navigable channel is, in some places, quite separate from the natural course of the river. Sir Richard Weston, who lived near Guildford, has the credit of introducing locks into Surrey. A Bill was passed in 1651 for the extension of the navigation up to Guildford, but the work was long in progress, not being completed till towards the close of the seventeenth century. In 1760, the Wey navigation was extended to Godalming, and there is no doubt that it added to the importance of this town and Guildford.

The Basingstoke Canal enters Surrey from Hampshire near Dradbrook. It turns northward to Frimley, where it changes its course in a north-easterly direction past Pirbright to the River Wey, which it enters a little below Byfleet.

The Surrey Canal was constructed in 1801 and runs from Camberwell to the Surrey Docks, whence it communicates with the Thames. The Croydon Canal was made at the same time. It began to the north of Croydon, and proceeded northwards to join the Surrey Canal at Deptford. Since 1836, when it was bought by a railway company, it has been partly used as the line of the South-Eastern, and London, Brighton, and South Coast Railways.

Another canal, now no longer used, was completed in the early years of the nineteenth century. This was the Surrey and Sussex Canal, which formed a junction between the Sussex Arun and the Surrey Wey. When the railways came, this canal was not needed, and now it is in many places filled up, while in others it is merely "a neglected

·THE·WEY·
·GODALMING·

ditch overgrown with reeds and water-lilies, haunted by the kingfisher and the moorhen."

Surrey was not only the first county to have a canalised river, but it also had the first railway in England. We should now call it a tramway and it was drawn by horses. It ran from Croydon to Wandsworth, and conveyed lime and building materials. Afterwards it was extended to Merstham, but it does not seem to have prospered.

The traction power of horses had to yield to steam power, and the first steam railway was from London to Greenwich in 1838. This was a very expensive line, for much of it was on brick arches. However, railways were destined to become the great means of communication, and there was no lack of money for their construction. In 1839, the Croydon line was finished, and in 1849 the line to Brighton was opened. The South-Western Railway carried their line from Woking to Guildford in 1845, then to Godalming in 1849, and finally to Portsmouth in 1859. The South-Eastern Railway from Redhill to Dorking, Guildford, and Reading was opened in 1849. Many other branches of these great railways have since been opened, and Surrey is now admirably served with the best railway communication.

23. Administration and Divisions—Ancient and Modern.

Before we consider the present administration of affairs in the county of Surrey, it will be well for us to get a clear idea of the ancient forms of government. It must first of all be clearly understood that many changes have been introduced into our parochial and county government since the time of our Saxon forefathers; but it has been the great care of most of our rulers to graft, as it were, new ideas on to the old English institutions. Thus it comes about that although we have improved methods of government to suit modern ideas, we can trace back, for a thousand years or more, many of our present institutions.

The government of a county, or shire, in early English times was partly central, from the county town, and partly local from the hundred, or parish. The chief court of the county of Surrey was, in the earliest times, the Shire-moot, which met twice a year. Its principal officers were the Ealdorman and the Sheriff, the last of whom was appointed by the King. In Saxon times, each county was divided into Hundreds, or Lathes, or Wapentakes. Surrey was divided into fourteen hundreds, and it is probable that, at first, each of these divisions contained one hundred free families. Each hundred had its own court, the hundred-moot, which met every month for business.

Each hundred was divided into townships, or parishes, as we now call them. Each township had its own assembly or *gemot*, where every freeman could appear. This gemot, or town-moot, made laws for the township, and appointed officers to enforce these *by*-laws, or laws of the town. The officers of the town-moot were the reeve and the tithing-man, who was the constable, and corresponded to our policeman. The court of the township was held whenever necessary, and the reeve was the president or chairman.

Besides these courts of the shire, of the hundred, and of the township, there were also courts of the manor, as the separate holdings of land were called. The manors were of different sizes, sometimes they were as large as the township, and sometimes they were parts of the township. The manors were held by their owners, or lords of the manor, as they were called, on various conditions. For example, they had to render service or homage to the King, and were allowed to sub-let their manors. The manor-courts were of various names, such as court-leet, court-baron, and customary court. In these courts, the lord and his tenants met, and settled the affairs belonging to the manor, such as those relating to the common fields, the rights of enclosure, and the holding of fairs and markets. These manor-courts are still held in many parts of Surrey, but they have long since lost the importance they once had; and we only refer to them here to show how interesting it is to remember that they are survivals dating back for more than one thousand years.

We are now in a position to consider the present form of administration of public affairs in the county of Surrey. The chief officers in the county are the Lord Lieutenant and the High Sheriff. The former is generally a nobleman, or a large landowner, and is appointed by the Crown. The Sheriff is chosen every year on the morrow of St Martin's Day, November 12.

The County Council now conducts the chief business of the county of Surrey. The County Hall was built at Kingston in 1894, and there all the meetings for the transaction of county business are held. Guildford, the

Abbot's Hospital, Guildford

old county town, was not chosen, as it was considered to be too far from London for this purpose. The County Council of Surrey consists of 19 Aldermen and 59 Councillors. The latter are elected to their office, while the former are co-opted by the Councillors for a term of years. The County Council corresponds to the ancient Shire-moot, and represents the central form of county government which was started in 1888.

For local government in towns and parishes, another Act was passed in 1894, when new names were given to the local governing bodies which had previously been known as vestries, local boards, highway boards, etc. In the large parishes, the chief governing bodies are now called District Councils, and of these there are 19 in Surrey. The smaller parishes have Parish Councils, or Parish Meetings. But whether District Councils or Parish Councils, they represent the old Town-moots, and the members are chosen by the people to manage the affairs of the district, and generally to advance its interests.

There are some towns in Surrey that have different and larger powers of government than the parishes. These larger towns are called Boroughs and are as follows: Croydon, Godalming, Guildford, Kingston, Reigate, Richmond, and Wimbledon. Croydon is the largest and most important, and has the dignity of a County Borough, with the power of a County Council.

Surrey has also 12 Poor Law Unions, each of which has a Board of Guardians, whose duty it is to manage the workhouses, and appoint relieving officers and others to carry on the work of keeping the poor and the aged.

For purposes of administering justice, Surrey has two Quarter Sessions that meet at Croydon and Guildford; and 11 Petty Sessional Divisions, each having magistrates or justices of the peace, whose duty it is to try cases and punish petty offences against the law. The boroughs of Godalming, Kingston, Reigate, and Richmond have each of them separate Commissions of the Peace.

If we go back to the earliest days of the Saxon conquest, we shall find that the Church existed as a body before the State; and its mode of government is much the same to-day as it was in those far-off times. Our land was divided into dioceses, over which were placed bishops. The northern dioceses and bishops were placed under the care of the Archbishop of York; while the southern dioceses and bishops were under the authority of the Archbishop of Canterbury. Surrey belonged to the diocese of Winchester, of which see it long formed an archdeaconry. In modern times, however, many ecclesiastical changes have been made, and now Surrey is in the sees of Winchester, Southwark, Rochester, and Canterbury. Each archdeaconry was divided into rural-deaneries and ecclesiastical parishes. At one time the ecclesiastical parish was the same as the civil parish, but now while there are 144 civil parishes in Surrey, there are many more ecclesiastical parishes.

The education affairs of Surrey are managed by Education Committees in the larger towns and parishes. The Surrey County Council has appointed a County Education Committee, and this has the control of secondary and elementary education in the rest of the county.

Charterhouse School, Godalming

Surrey is represented in the House of Commons by six members from the following divisions: Chertsey, Guildford, Reigate, Epsom, Kingston, and Wimbledon. The Borough of Croydon also returns a member, and several of the boroughs in the County of London are also represented. It is interesting to note that the old parliamentary boroughs before the Reform Act of 1832 were Guildford, Southwark, Reigate, Bletchingley, Haslemere, and Gatton.

24. The Roll of Honour of the County.

A great writer of old urged us to praise famous men, and there is no doubt that it is both profitable and interesting to know something of the celebrated men who have shed lustre on the place of their birth or residence. Let us consider in this chapter some of the worthies of Surrey, whose names are familiar in our mouths as household words.

As Surrey is a metropolitan county we are not surprised to find that many of our monarchs have resided within its borders. There was a royal palace at Richmond, or Sheen, as it was formerly called: Edward III lived there, Henry V built it afresh, and Henry VII restored it. The palace of Sheen was the residence of most of our Tudor sovereigns; Henry VIII hunted in Richmond Park; and Elizabeth, the greatest of the Tudors, passed away in this palace, which no longer stands. Henry VIII had another house at Cuddington, and so splendid was it that men called it "Nonsuch," because there was no other like it. There was another royal house at Kew, where George III spent much time when he was free from the cares of State.

The official residence of the Archbishops of Canterbury is Lambeth Palace, and down to quite recent times there was another house belonging to the See of Canterbury at Addington. Hence we find a number of the great Archbishops, besides other dignitaries of the Church, living in Surrey. Becket went to school at Merton, where he was educated by Robert Bayle, the first prior of that abbey. Walter de Merton was born and educated at Merton. He subsequently became Bishop of Rochester, and founded Merton College, Oxford. Cardinal Wolsey had a house at Esher, and thither he retired in 1529, when he was deprived of the Great Seal. Archbishop Whitgift, who held the See of Canterbury in Elizabeth's reign, has left a fine memorial of his generosity, for he built a hospital at Croydon as a refuge for poor men and women. A school was attached to this charity, and both institutions are still flourishing. Archbishop Abbot held office from 1611 to 1633, and is remarkable from the fact that he raised himself from a lowly position to the highest in the Church. Abbot was born at Guildford, educated at the grammar school of his native town, and buried at Holy Trinity Church of the same place. Like Whitgift, he built some almshouses for old men and women, and these fine buildings occupy a commanding position in Guildford High Street. We must mention among the famous Surrey divines the

Lambeth Palace

honoured name of Wilberforce, who was Bishop of Winchester in the middle portion of Queen Victoria's reign. He was a popular bishop and an orator. Wilberforce met his death at Evershed Rough, near Abinger, owing to a fall from his horse. A granite cross commemorates this sad event, which occurred on July 19th, 1873.

Among the statesmen who have lived in Surrey we will mention Thomas Cromwell, Earl of Essex, Sir William Temple, and Speaker Onslow. Thomas Cromwell was by far the greatest of them, and has left his name engraved in the annals of our land. The son of a blacksmith at Putney, Cromwell worked his way to be Henry VIII's chief adviser, for he succeeded Wolsey, his master, in that high position. It was Cromwell who was Henry's trusted agent in closing the monasteries, the spoils from which went to enrich the King and his favoured nobles. Arthur Onslow was Speaker of the House of Commons for thirty-three years, and during that time, it is said, "he filled the chair with higher merit probably than anyone either before or after him, with unquestioned impartiality, dignity, and courtesy." Onslow lived from 1691 to 1768, and was Recorder and M.P. for Guildford.

Men of action are conspicuous among the worthies of Surrey. Perhaps the first worthy of mention is Lord Howard of Effingham, who commanded the English attack on the Spanish Armada. He died at Haling House, near Croydon, and was buried in a vault beneath the chancel of Reigate Church, in 1624. Lord Clive, who won India for England, built a fine house at Claremont, where he lived on his return from the East till his death in 1774. Lord Rodney, the "breaker of the line," one of our great admirals, was buried at Walton in 1792; and Lord Nelson, the greatest of all our seamen, lived at Merton Place from 1801 to 1803. Nelson's house no longer stands, and the grounds are covered with small buildings.

Merton Place, Lord Nelson's Villa

Surrey is perhaps the most beautiful county in England, and has been the home of many of our sweet singers. One of our earliest poets, John Gower—"Moral" Gower, as he was called—lies buried in Southwark Cathedral, where there is an effigy to his memory. Chaucer has made Southwark famous for all time, as it was at the Tabard Inn there that his company of pilgrims started on their journey to Canterbury. Shakespeare, Massinger, Beaumont, Fletcher, and Edward Alleyn were either dramatists or actors who won fame at some of the theatres on Bankside, Southwark; and in the cathedral of that borough there are memorials to them. Edward Alleyn, an actor, has left his mark in the history of Surrey, for he founded the College of God's Gift at Dulwich in 1613, now one of our great public schools.

Sir John Denham, who lived at Egham, made himself famous by his poem *Cooper's Hill*, which was praised both by Dryden and Pope. He was called "Majestic Denham," and certainly his poetical description of the scenery of this part of the Thames Valley is very fine. James Thomson, the author of *The Seasons*, lived and died at Richmond, and in *Summer*, one of the portions of his long poem, he gives a picturesque description of the view from Richmond Hill. The last four lines are worth quoting:

> "Heavens! what a goodly prospect spreads around,
> Of hills, and dales, and woods, and lawns, and spires,
> And glittering towns, and gilded streams, till all
> The stretching landscape into smoke decays."

Among our later poets, Robert Browning was born at Camberwell, in 1812; and Tennyson lived at Aldworth House, not far from Hindhead. This was his summer residence, and here he passed away. Tennyson thus describes the view from his residence:—

> "You came, and look'd, and lov'd the view,
> Long known and lov'd by me;
> Green Sussex fading into blue,
> With one grey glimpse of sea."

John Evelyn Esq.r

Pour Mons.r de La Quintenay.
recommendé a Mon.r [illegible]

Evelyn.

Evelyn.

His Autograph from an Original in the Possession of
John Thane

After the poets, it is perhaps well to turn to the men of letters who have made Surrey famous. John Evelyn, author of the *Diary*, must come first, for not only was he a good writer, but he did much for Surrey in the way of tree-planting. He built part of Wotton House, and laid out the grounds. In the neighbourhood of Wotton and Leith Hill there are a number of fir-trees and larches, and it is owing to Evelyn that these conifers, afterwards widely cultivated in England, were introduced into Surrey. It has been well said that Evelyn's tomb may be seen in Wotton Church, but his real monument is in the ornamental woods of Surrey.

Dean Swift wrote many of his works at Farnham, and Gibbon, the historian of the *Decline and Fall of the Roman Empire*, was a native of Putney. William Cobbett, in many ways one of the most interesting of Surrey's worthies, was born at the "Jolly Farmer" public house at Farnham, in 1762, and he was buried in the parish church of his native place in 1835. Cobbett loved Surrey with all his heart, and wrote with feeling of its beautiful scenery. His writings appealed to the men of his day, and are marked by strong convictions expressed in simple, nervous English. Cobbett's *Rural Rides* gives us a graphic picture of the state of the land in the eastern and south-eastern counties; and his *Advice to Young Men* was once exceedingly popular. John Ruskin, one of the greatest prose writers of his day lived in Croydon, and with him we must close our record of the men of letters.

When we turn to the men of science who lived in Surrey, we must be content to mention John Tyndall, a natural philosopher, who had the gift of making science attractive by his lectures and writings, and lived for many years at Hindhead. On the summit of that famous hill, he built himself a house, "foursquare to the

William Cobbett

winds" and protected by a huge screen from the overlooking of the curious. There he wrote many of his great works, and there he died in 1893.

Here we must end our review of Surrey worthies. The Roll of Honour is a long one, and we have had space only to deal with the more commanding names. Many men of great merit have helped to make Surrey famous, and it should be the pride of Surrey people to hold their names in memory. There are few places where "some village Hampden," or "some mute inglorious Milton," or "some Cromwell," has not flourished, and it is always pleasant to connect such men with our own town or parish.

25. THE CHIEF TOWNS AND VILLAGES OF SURREY.

(The figures in brackets after each name give the population in 1901, and those at the end of each section are references to the pages in the text.)

Addington (642) is a small village, three miles south-east of Croydon. Addington Park was one of the residences of the Archbishops of Canterbury, but was sold after the death of Archbishop Benson. (pp. 85, 102, 124.)

Albury (1224), five miles south-east of Guildford, is of great interest, as there are traces of barrows and tumuli in the neighbourhood, and large numbers of British and Roman coins have been found.

Banstead (5624) is beautifully situated on the chalk downs, at an elevation of about 520 feet. It has an Early English Church, with a lofty spire that serves as a landmark for some distance round.

Barnes (10,047) is a parish lying about the common of the same name to the south of Hammersmith. It was the residence of Fielding the novelist, Cowley the poet, and Handel the musician.

Battersea (168,907) is a thickly populated parish on the south of the Thames, opposite Chelsea, with which it is connected by two bridges. Battersea Park is an extensive public ground, with many beautiful trees and plants. Battersea and Clapham form a parliamentary borough. (pp. 33, 71.)

Beddington (3844), on the Wandle, possesses one of the most beautiful churches in Surrey, with many interesting features—a Norman font, an Elizabethan pulpit, and carved stalls.

Bermondsey (130,760) is part of the parliamentary borough of Southwark. It is now a crowded and unpleasant district, and is the chief seat of the tanning industry. The most interesting feature of Bermondsey in olden times was its Abbey, founded about 1082; but of this fine building nothing remains but portions of the east gateway. (pp. 66, 98.)

Betchworth (1789), a pretty village on the Mole, is situated between Dorking and Reigate. It has a Norman church with later additions, and there are remains of its castle. (p. 102.)

Bletchingley (2128), once a parliamentary borough and market town, is now a picturesque old village. Its castle, so famous in the Barons' Wars, stood on a hill at the west end of the town. A portion of the wall of the keep and the outer and inner ditches alone remain. (pp. 85, 102, 123.)

Brixton (9492) is now in the County of London and forms part of the parish of Lambeth. It is densely populated, but quite uninteresting.

Brookwood is famous for the London Necropolis, a very beautiful cemetery formed out of the wild heath-land. The grounds are well planned, and in summer bright with flowering shrubs and other blossoms.

Camberwell (259,339) is one of the populous southern suburbs of London. It was once the stopping-place of the pilgrims on their way to Canterbury. Sir Christopher Wren resided here. (pp. 71, 128.)

Carshalton (6748) is a pretty town on the Wandle. Near the church is Anne Boleyn's Well, which is reported to have sprung forth from a stroke of her horse's hoof. In the neighbourhood are herb farms, market gardens, and several mills. The local pronunciation of the name is Casehawton. (p. 71.)

Caterham (9486) is a rapidly growing parish in the valley of that name.

Cheam (3404) lies one mile west of Sutton. In the village is an ancient timber house called the Council House, where, tradition asserts, Queen Elizabeth held her Councils when staying at Nonsuch Palace.

Chertsey (12,762) is a pleasant riverside town, and a good centre for boating and fishing. Of the celebrated Abbey founded in 666 nothing remains but part of a wall. Chertsey Bridge, a picturesque structure, spans the Thames and gives connection with Middlesex. Among the famous residents of Chertsey were Abraham Cowley, the poet, who lived and died in the Porch House, and Charles James Fox, the great Whig orator, who lived at St Anne's Hill. (pp. 23, 27, 78, 95, 123.)

Chiddingfold (1548) is a delightful village, but at one time a manufacturing centre with iron-smelting and glass-making industries. (pp. 65, 86, 104.)

Chobham (3186) on the Bourne, a feeder of the Wey, has an Early English church of some interest. In the vicinity lie Chobham Ridges, a line of heather-clad hills of great beauty. It has been said that "in August, when the heather is in bloom, one might fancy himself on a Scotch moor, miles away from civilisation." (p. 12.)

Clapham (51,361) is a large parish and suburb in the south-west of London, with a great centre of railway activity at Clapham Junction, through which upwards of 1200 trains pass daily. Clapham Common is a well-kept playground, 200 acres in extent. Clapham was the home of many great men; Macaulay, Wilberforce, and others of the "Clapham Sect" made this place famous.

Cobham (3901), on the Mole, seven miles south-west of Kingston, is in the midst of a beautiful country of pine-woods and commons. There used to be ironworks on the Mole, but they are now used as saw-mills.

Crowhurst Manor House

Cranley, or **Cranleigh** (2709) is a pleasant village scattered about an old green. The church is old and has been carefully restored.

Crowhurst (243) is in the Weald, two miles south-east of Godstone station. There is an interesting old church, and in the churchyard is a yew-tree said to be 1500 years old. (pp. 43, 74.)

Croydon (133,895), a large and growing town on the Wandle, is nine miles south-east of London Bridge. It is a place of great antiquity, and was held by the Archbishop of Canterbury before the Domesday Survey. Lanfranc built a palace here, which was used by the Archbishops down to the eighteenth century. During the Barons' War, in 1264, the Londoners who had retreated from Lewes were attacked by the Royal troops and defeated with heavy loss. The plague raged in Croydon in 1603-4, and again in 1665-6, causing many deaths. Until quite recently, there were many picturesque old houses in the High Street, but all of them have now disappeared. The chief buildings of interest are the Archbishop's Palace, the Whitgift Hospital, the Parish Church, and the new Municipal Buildings. The Palace is being restored and will be devoted to religious purposes. The Whitgift Hospital is a noble foundation with an endowment yielding £4000 a year. It maintains thirty-nine poor persons, and supports two schools. The Parish Church is the finest church in Surrey, and has a monument to Archbishop Whitgift. The Municipal Buildings, consisting of Town Hall, Courts of Justice, Corn Exchange, and Free Library were opened by the King and Queen in 1896. Brewing and shoe-making are among the industries of Croydon, and there is also an important bell-foundry. (pp. 7, 14, 54, 68, 84, 91, 120, 124, 130.)

Cuddington (774) is a small parish, two miles north-east of Ewell. It is memorable for having contained Henry VIII's magnificent palace of Nonsuch. This fine residence afterwards passed to the Earl of Arundel, who sold it to Queen Elizabeth. Here she received the Earl of Leicester on his return from Ireland in 1599, and marked her displeasure of her favourite. (p. 123.)

Ditton, Long (2080), one mile south of Kingston, has a modern church (1880), which replaced one of older design. (p. 28.)

Ditton, Thames (4986) is a prettily situated village on the Thames. Its Perpendicular church has some interesting monuments and brasses. (pp. 19, 28.)

Dorking (11,410) is beautifully situated on a tributary of the Mole, and lies in the Weald between the slopes of Box Hill and Leith Hill. The Roman road, Stone Street, passed through the old churchyard, and Roman remains have been found. There are many good residences in the neighbourhood of Dorking; but perhaps Deepdene, with its beautiful grounds, is the best known. (pp. 14, 17, 70, 84, 112, 113.)

Dulwich (97,369), a suburb of London and part of Camberwell, is chiefly famous for its College, founded by Alleyn in the reign of Elizabeth. The "College of God's Gift," as the foundation is called, has a Picture Gallery, with a fine collection of pictures. (p. 128.)

Egham (11,895) is a large parish of considerable interest both in ancient and modern times. It includes a portion of Windsor Great Park, and Virginia Water, while Runnymede is to the north, and Cooper's Hill to the north-west of the town. The Royal Holloway College on Egham Hill was founded, at a cost of £600,000, for the higher education of women.

Epsom (10,815), a picturesque old country town, was once famous for its Wells, situated on the Common. Great numbers of people visited Epsom in Stuart times, owing to the medicinal spring. Epsom has been a favourite resort for horse-racing since the time of James I; but the two great annual races, the Derby and the Oaks, date from the latter part of the eighteenth century. (pp. 9, 14, 39, 123.)

Esher (2423) is an old village situated on high ground on the Portsmouth Road. Esher Place is a modern mansion, which takes the place of the old palace of the Bishops of Winchester. Claremont Park lies to the south of Esher, and the fine house was built in the reign of Queen Anne. Here it was that Lord Clive ended his days. Since the beginning of the nineteenth century it has belonged to the Royal Family. (pp. 19, 49, 90.)

Ewell (3338) stands at the head of Hog's Mill River, and is chiefly of note from the remains of King Henry VIII's palace of Nonsuch. (p. 23.)

Farnham (14,582) is an old town on the Wey, and long noted for its castle, the residence of the Bishops of Winchester. The castle is on high ground in the corner of a beautiful park and dates from about 1136. Hops are grown largely in Farnham, and form an important part of the trade of this ancient town. William Cobbett was born in Farnham at the "Jolly Farmer" inn. (pp. 14, 17, 20, 39, 45, 59, 78, 79, 82, 100, 106, 130.)

Frensham (2103) is a village prettily situated on the Wey and in the midst of the beautiful heather-district on the borders of Hampshire. There is an extensive common, and several large ponds cover an area of 300 acres.

Frimley (8409) is a picturesque village at the foot of Chobham Ridges, in the valley of the Blackwater. Here is the Royal Albert Orphan Asylum, which maintains upwards of 200 children.

Godalming (10,515), an ancient town with many picturesque old houses, is situated on the Wey. The town has industries connected with paper-making, tanning, and hosiery. The most notable modern building is the Charterhouse School, which was removed from London in 1872. (pp. 21, 64, 66, 69, 92, 105, 114, 120.)

Godstone (2800) is a pleasant village at the head-waters of the Bourne.

Guildford (8417), the capital of Surrey, is beautifully situated on the River Wey, which flows through a gap in the Downs. The High Street mounts one of the steep hills, and has many picturesque timber houses and famous buildings. The history of the town from the time of King Alfred is of considerable interest; and the

Castle, of which only the keep and bits of the wall remain, was once of great importance in the county. Of the three churches, the most interesting is St Mary's, a Norman building with later additions. The Town Hall, in the High Street, dates from 1683 and contains some Stuart portraits. Abbot's Hospital, founded in 1619, is built of red brick in late Tudor style, and is, perhaps, the most picturesque building in the High Street. (pp. 14, 17, 21, 39, 49, 63, 65, 69, 79, 86, 91, 92, 97, 100, 105, 107, 114, 118, 120, 123, 124.)

Ham (1460) is a village, with a fine common, between Richmond Park and the Thames. Ham House, a Jacobean mansion, has fine collections of books, pictures, furniture, and art treasures. (p. 59.)

Haslemere (2614) is a small market town at the south-west of the county, on the borders of Sussex and Hampshire. North of the town is a charming stretch of moorland that terminates in the well-known hill, Hindhead, 895 feet high. The whole neighbourhood is exceedingly beautiful, and is becoming a popular centre for people seeking country houses at a fair distance from London. (pp. 7, 12, 49, 51, 68, 72, 90, 104, 123.)

Kew (2699) is a well-known parish on the Thames, and the village clusters round the common. Kew Bridge, a very fine modern structure, connects Surrey and Middlesex. The church of red brick has monuments to Gainsborough, Sir William Hooker, and the Duke and Duchess of Cambridge. Kew Palace dates from the time of Queen Elizabeth, and Kew House was formerly the residence of George III. The Royal Botanic Gardens, the most beautiful in England, were founded in 1760 by command of the King, and were presented to the nation in 1840 by Queen Victoria. Sir William Hooker was appointed the first Director, and he and his successors have made the gardens the most famous of their kind in the world. The pleasure grounds are well laid out, and the conservatories, the great palm house, the water-lily house, and the tropical fern house are of the greatest interest. The Pagoda, 163 feet high, the Temple of Victory, and the Flag Staff, 159 feet high, formed of the trunk of a Douglas pine, are all features of interest in the grounds. (pp. 32, 51, 91, 124.)

Kingston-on-Thames (34,375) is probably the oldest town in Surrey, and has many associations connected with Roman and Saxon times. The name means "King's town," and was probably given from the fact that many of the Saxon Kings were crowned

BRIDGE AT LEATHERHEAD

here. The stone on which they are supposed to have stood, or sat, at their coronation, is carefully kept in the Market Place. The County Hall, where the Surrey County Council meets, is situated in the town. (pp. 23, 28, 80, 85, 88, 91, 102, 118, 120, 123.)

Lambeth (301,895), on the Thames, is a parliamentary borough opposite Westminster. It is chiefly famous for Lambeth Palace, which has been the town residence of the Archbishops of

Canterbury since the twelfth century. The palace is mainly of red brick, and has been closely associated with the history of the Church of England for nearly 800 years. The Chapel, the Library, the Gateway Tower, and the Lollard's Tower are the chief features of interest in this historic building. St Thomas's Hospital is in Lambeth. Among the industries are those connected with pottery, glass, soap, white lead, etc. (pp. 33, 53, 106, 124.)

THE·VALE·OF·MICKLEHAM·

Leatherhead (4694) is a pretty little town on the Mole, which is crossed by a handsome bridge. (pp. 14, 19, 39, 90.)

Merton (4510) is a very ancient town, once celebrated in English history. Its great priory was a most important monastic foundation, and at the priory school Thomas à Becket and Walter de Merton were educated. Nelson lived at Merton, and here William Morris established some tapestry works. (pp. 23, 78, 97, 124, 127.)

Mickleham (750) is a village most beautifully situated in the Vale of Mickleham, on the banks of the Mole, between Leatherhead and Dorking. (pp. 19, 90.)

Mitcham (14,903), on the Wandle, is famous for the cultivation of lavender and other aromatic herbs. (pp. 59, 68.)

Molesey, East (5119) stands at the point where the river Mole falls into the Thames. (p. 28.)

Mortlake (7774), two miles east of Richmond, has been for some years the winning post in the Oxford and Cambridge Boat Race. (pp. 9, 32, 53, 59.)

Newington (121,863) is a parliamentary borough on the south of the Thames. The place was called Newington Butts in the fourteenth century, when butts for archers to shoot at were placed here. The district is now devoid of interest and is only remarkable for its dense population.

Norwood (21,417) is a large suburban district lying between Camberwell and Croydon. The neighbourhood is very hilly, and has many large and attractive residences. The Crystal Palace is partly in Norwood and partly in Sydenham, a Kentish parish. Norwood is remarkable for its large number of public institutions, of which the Royal Normal College for the blind is, perhaps, the best known.

Nutfield (1860), a pleasant village two miles east of Redhill, is famous for fuller's earth, which is dug in the neighbourhood in large quantities. (pp. 39, 71.)

Ockley (565) was formerly known as Stone Street from its position on the old Roman road. It is supposed that a battle was fought here in 851, when Ethelwulf defeated the Danes. (pp. 77, 102.)

Petersham (589) is a pleasantly situated old village at the foot of Richmond Hill. Petersham Park forms part of Richmond Park. (p. 91.)

Putney (24,139) is an extensive suburban district along the south of the Thames, opposite Fulham. It is a popular rowing centre, but there is not much of interest in the place. It was the birthplace of Thomas Cromwell, Earl of Essex, and of Gibbon, the historian. (pp. 9, 32, 110, 130.)

Reigate (25,993) is a picturesque old town at the south foot of the North Downs. It is styled Cherchefelle in Domesday Book, and its present name, which means Ridgegate, or passage through the ridge of the North Downs, was not used till the thirteenth century. Reigate was the seat of de Warrenne's Castle, the grounds of which now belong to the town. The parish church, dating from the twelfth century, is one of the finest in Surrey. (pp. 17, 79, 82, 97, 101, 120, 123.)

Richmond (25,577) is a municipal borough about 10 miles from London. Previously to the reign of Henry VII, its name was Sheen. Many monarchs lived at Richmond Palace, and it was here that Queen Elizabeth died in 1603. Nothing now remains of the palace but the gateway in Richmond Green. Richmond Park was added to the royal domain by Charles I in 1637. It has an extent of over 2200 acres, and is well stocked with deer. Richmond is a favourite residential resort of Londoners, and has been the home of many great men, *e.g.* Thomson the poet, Edmund Kean the actor, and Gilbert Wakefield. The Thames is crossed at Richmond by a stone bridge of five arches. The view from Richmond Hill is justly famous, and has inspired poets to sing its praises. (pp. 30, 37, 45, 49, 54, 68, 120, 123, 128.)

Ripley (2201) is a modern parish formed out of that of Send. A broad street runs through the pretty village, which is a favourite resort of cyclists and motorists.

Rotherhithe (38,460) was once part of the royal manor of Bermondsey. The Commercial Docks are here, and are entered by the Grand Surrey Canal.

Shere (2184) on the Tillingbourne, is one of the most beautiful of Surrey villages, and is much visited by artists.

Southwark (206,180) is a parliamentary borough situated on the south side of the Thames, opposite the City of London. Now it is a most densely populated district, but its historical interest is very great. Its early importance arose from the fact that it was on the line of Watling Street, and near the great crossing place on the southern bank of the Thames. Its priory was closed at the Reformation, but the priory church was bought by the people of Southwark and now forms the Cathedral Church of the see. The Tabard Inn, connected with Chaucer's *Canterbury Tales*, and some of the theatres, were associated with Shakespeare and other Elizabethan actors and dramatists. Guy's Hospital, an old foundation, is in this borough, and there is a large market. This populous district has many manufactures and industries, *e.g.* glass-making, mat-making, brewing, and tanning. (pp. 33, 66, 80, 85, 89, 92, 128.)

Streatham (71,658), a large and populous suburb to the south of Brixton, is said to owe its name to its position on a Roman road or street. (p. 37.)

Surbiton (11,981), a suburb of Kingston, was the scene of the last struggle of the Royalists on behalf of Charles I. (p. 7.)

Sutton (17,223) is an extensive parish about four miles west of Croydon. It was formerly the first stage on the coach road from London to Brighton. (pp. 7, 71.)

Tooting (16,473) is a large suburban district. Its common of 144 acres is very beautiful with its trees and gorse. (p. 23.)

Walton-on-Thames (10,829) has a bridge across the Thames to Halliford. It was near here that Caesar is supposed to have crossed from the south to the district north of the Thames. (pp. 21, 28, 74, 127.)

Wandsworth (68,403), a large suburb in the south-west of London, takes its name from the river Wandle, which here falls into the Thames. In the seventeenth century many French Huguenots settled here and introduced various manufactures. (pp. 23, 33, 53, 68.)

Weybridge (5329) is on the Wey, where that river enters the Thames. It is on the borders of the pine district, and near some of the most beautiful scenery in Surrey. (pp. 28, 85.)

Wimbledon (41,652) is a municipal borough whose history goes back to the earliest times, foı it was the scene of a battle in 568 between Ethelbert, King of Kent, and Cealwin, King of Wessex. Wimbledon Common of 1000 acres was once the "camp" of the volunteers. In 1890 the camp was removed to Bisley. (pp. 7, 54, 76, 81, 120, 123.)

Woking (16,244), on the Wey, is a large parish with paper-mills and extensive nursery-grounds. (p. 21.)

Wotton (608), three miles south-west of Dorking, was the home of the Evelyns. Here John Evelyn was born and educated, and he was buried in the parish church. (p. 130.)

Fig. 1. Diagram showing the area of Surrey compared with that of England and Wales

Fig. 2. Diagram showing the population of the administrative county of Surrey compared with that of England and Wales

Fig. 3. Diagram showing the population of Surrey, including the portion in the County of London, compared with that of England and Wales

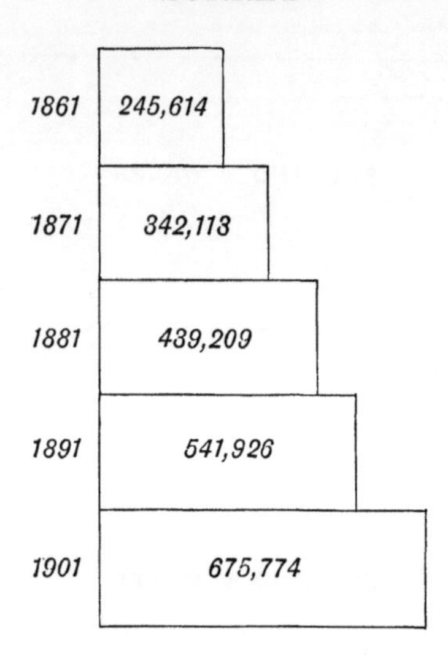

Fig. 4. Diagram showing the increase in the population of Surrey from 1861 to 1901

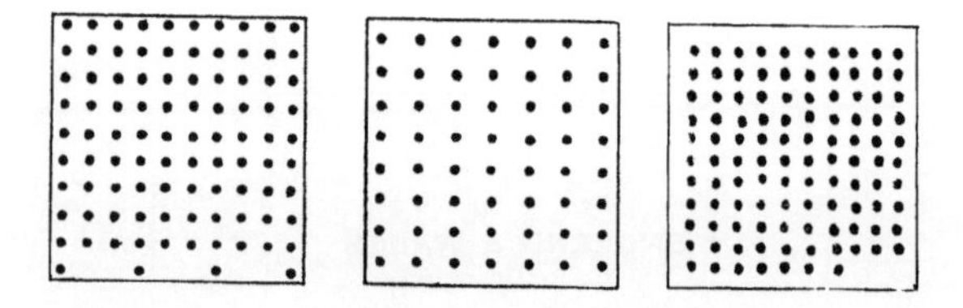

Surrey 936 England and Wales 558 Lancashire 1070

Fig. 5. These diagrams show the comparative density of population to the square mile in Surrey, England and Wales, and Lancashire, 1 dot = 10 persons

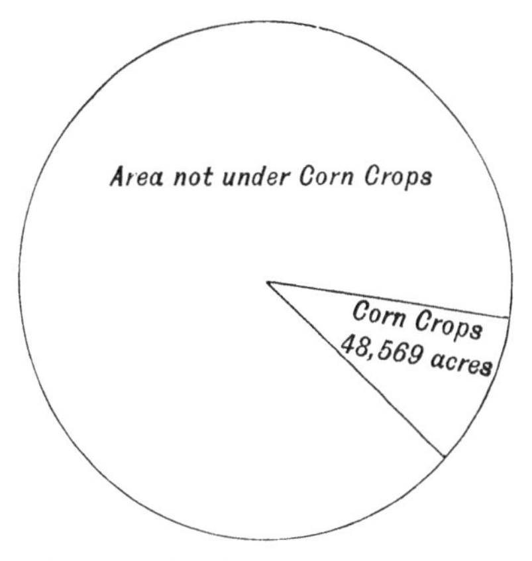

Fig. 6. Diagram showing the proportionate area of Corn Crops to total area of County

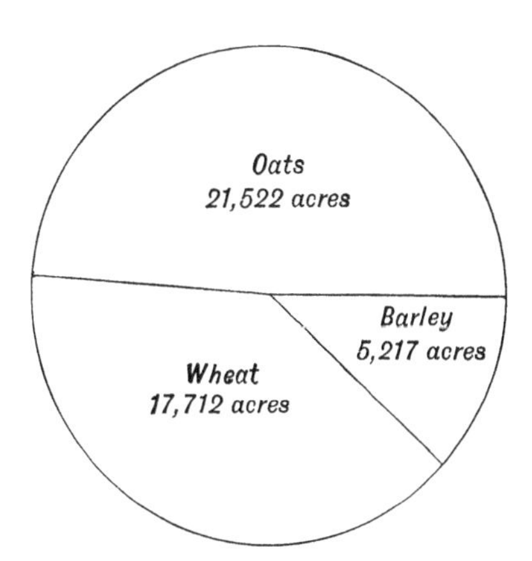

Fig. 7. Diagram showing the proportionate area of Oats, Wheat, and Barley in Surrey

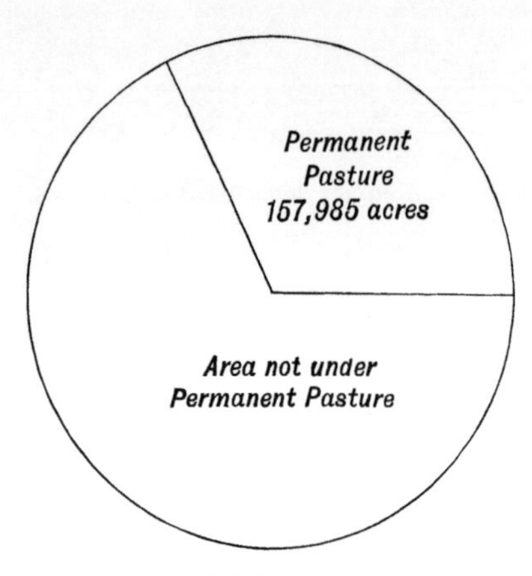

Fig. 8. Diagram showing the proportionate area of Permanent Pasture to total area of the County

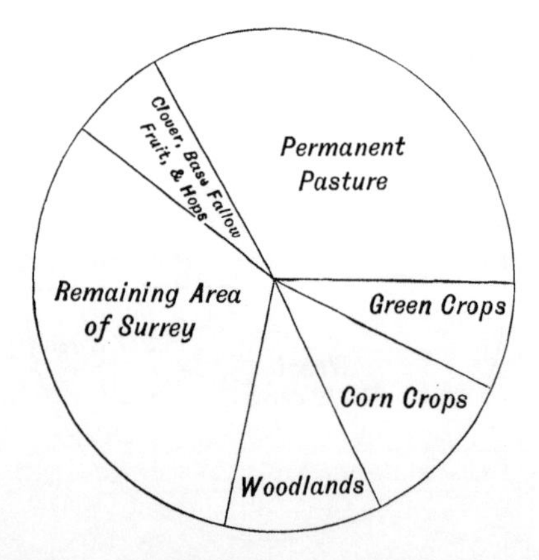

Fig. 9. Diagram showing the proportionate areas of Portions under Cultivation and Not under Cultivation

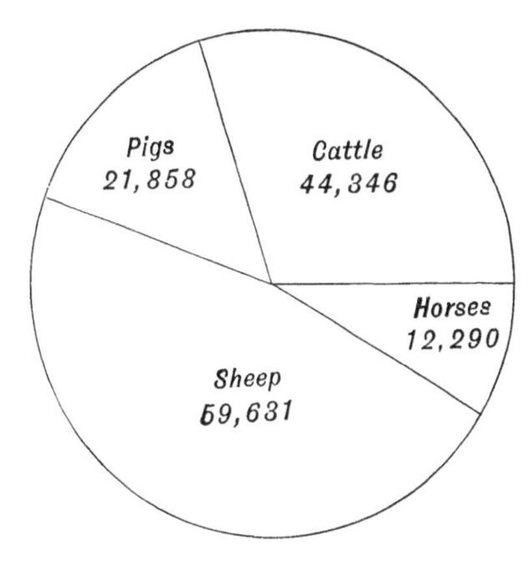

Fig. 10. Diagram showing the comparative number of Sheep, Cattle, Horses, and Pigs in Surrey

www.ingramcontent.com/pod-product-compliance
Ingram Content Group UK Ltd.
Pitfield, Milton Keynes, MK11 3LW, UK
UKHW042145280225
455719UK00001B/117

9 781107 675988